Artificial Intelligence, Simulation and Society

Series Editor

Petra Ahrweiler, TISSS Lab, Institute for Sociology, Johannes Gutenberg University of Mainz, Mainz, Germany

This book series brings into its fold key and emerging topics on the interactions between growing artificial intelligence technologies and their social impacts. It addresses various aspects of the relationship between AI, simulation, and society and provides insights into their intersections and stimulates discussions on the opportunities and challenges they present. The series is multi- and transdisciplinary in scope, and dynamic. It invites academic contributed volumes and monographs, but also more popular work suitable for lay readership, and innovatively includes some science fiction to initiate readers into the scope and aims of this novel series.

The specific themes and topics covered under the series are:

- **The ethical and societal implications of AI**: The series delves into the ethical considerations and societal impacts of AI technologies. It explores topics such as privacy, bias, job displacement, and the role of AI in shaping social structures from a social science point of view (sociological, political, economic, cultural, legal).
- **Simulation and modeling of social systems**: The series explores how simulation techniques are used to model and understand complex social systems and create artificial societies in silico. It covers topics such as social network analysis, agent-based modelling (ABM), and the simulation of collective behaviour.
- **AI and social simulation**: The series explores how AI technologies are used in social simulation, for example, modelling intelligent agents in agent architectures of ABM, or calibrating and validating models using intelligent data mining and analysis techniques.
- **AI and simulation in social philosophy**: It looks at how AI and simulation are depicted in social philosophy, for example, the role of AI and simulation in socio-technical evolution, the position of AI and simulation in Western rationalism, philosophical counter-designs of current developments, ontological and epistemological limitations and barriers of AI and simulation.
- **AI, simulation and society in fiction**: The series also innovatively examines the portrayal of AI and simulation in and as fiction, demonstrating how these themes reflect societal fears, aspirations, and ethical dilemmas. The series contains both original fiction and second-order analyses.
- **AI and simulation in entertainment:**It covers simulation techniques, combined with AI, that are used to create virtual worlds and characters that mimic human behaviour. Such simulations are used, for example, in video games, virtual reality experiences, and entertainment applications.
- **AI and simulation in various disciplines**: The series discusses the applications of AI and simulations that are/will be transforming various disciplines and domains such as healthcare (e.g. in medical diagnosis, drug discovery, and patient care), work (e.g. automation, Industry 4.0, workforce dynamics), or education (e.g. virtual reality, personalised learning systems, intelligent tutoring systems). It discusses the potential benefits and challenges of integrating these technologies into the conventional space.
- **AI, simulation, and policy**: The series analyses how AI and simulation techniques can inform the policy cycles. It discusses the use of predictive modelling, analysis of what-if scenarios, and decision support systems in shaping policies in various policy domains such as public policy, technology policy or environmental policy.

Alexandru Balasescu

Climate Change in the Age of Artificial Intelligence

Nature, Culture and the Politics of Technology

Springer

Alexandru Balasescu
Royal Roads University
Victoria, BC, Canada

ISSN 3004-9822 ISSN 3004-9830 (electronic)
Artificial Intelligence, Simulation and Society
ISBN 978-3-031-90041-9 ISBN 978-3-031-90042-6 (eBook)
https://doi.org/10.1007/978-3-031-90042-6

This Springer imprint is published by the registered company Springer Nature Switzerland AG
The registered company address is: Gewerbestrasse 11, 6330 Cham, Switzerland

To Zoë Balasescu and her future; for Ioana David, Marcella Blagoe, Aurelian Popescu, my mother Sanda, and the past.

Foreword

As I write this foreword for Dr Balasescu's detailed and important work, I find myself looking out over the seemingly desolate Arctic tundra, and thinking about thinking. If I look at the distant hills, bereft of trees or scrub as might be present farther south, as an artist, I might see unparalleled beauty; I might see with my eyes, my very human eyes, a mood or a feeling, possibly even a reflection of facets of myself, or of humanity itself. If I were to look at the same scene beyond my window as an economist (a discipline in which I am most certainly not any kind of expert), I might see a mineral-rich opportunity defined by numbers and formulae. If I were to look at the same scene, this time as a healthcare professional, I might see a remote and forbidding land in which fruit and vegetables cost so much that many here cannot afford a healthy diet. Which of these views is correct, and perhaps more importantly, what do we mean by "correct?"

At this moment, I am in Iqaluit, the capital city of the territory of Nunavut, with colleagues from the university in which I teach. We are here to listen to what the people who live here, particularly those in government, might want from higher education. It feels like important work, work that could have far-reaching implications to the people that live here, their livelihoods, and their governance. The very name: Nunavut, meaning "our land" in Inuktitut, is relevant to our work here. How might we modify the standard fare we offer to accommodate the real needs of the people who live in this unique place? And of course, government here, like government everywhere in my experience, is about detail, policy, management, those things we can measure, but perhaps not those things we know to be true, but cannot be measured. So in this moment I feel hope and frustration, both of which can be found on the pages of this book.

My work in the world, mostly in research and teaching, has focused on systems theory, and how we can better understand and work with human and organizational systems. Naturally, this leads to a somewhat tighter focus on leadership. But of course leadership, or more accurately our understanding of leadership, over the past few decades, has exploded into a vast and often controversial field of knowledge and practice. Many consultants, and indeed some of my colleagues, present elegant models that purport to explain and address the complexities of human behavior.

More often than not, these models emphasize the importance of measurement in any and every initiative from project management to day-to-day administration. Of course, measurement is important, even essential, but it is not the whole story, is never the whole story.

In project management, the Project Management Body of Knowledge, or PMBOK as it is more commonly known, divides the work of project management into the art and the science. My friend and colleague, Dr Dale Christenson, one of a handful of humans in the world with a doctorate in project management, and with whom I teach the subject, is an expert in the science of project management. Teaching the art of it falls to me. But because of our understanding of these topics, and perhaps our friendship as well, the two focuses blend into one, each indistinguishable from the other. When we say "science," we often think of the Natural Sciences. We think of formulae and measurement, isolating variables, and uncovering defensible knowledge. Certainly this is Dr Christenson's wheelhouse. But he also speaks of navigating change, a process in which one cannot isolate variables, and where each situation is unique. The art and the science blend together, and perhaps that is where the reality of project management might be briefly glimpsed. And it is this capacity to rise above one paradigm or perspective, to embrace art *and* science, that Dr Balasescu's writing inspires.

Similarly, in the work that I share with my dear friend Alec, teaching global leadership, he, an anthropologist, Dr Wanda Krause, a political scientist, Erin Dixon, an amazing Métis woman, and Community Difference-Maker, all teach the same thing, albeit from different perspectives, and using different lexica. The results of our different perspectives and knowledge coming together to inform such a complex field are extraordinary. Our topics and worldviews, each unique, come together to create a compelling whole.

If science is a discipline for building and organizing knowledge, then science is certainly what my global leadership colleagues and I have been doing these past years. But if science includes a predictably systematic approach, isolating variables so that all other things are equal, and involves measuring and testing hypotheses, then what we are doing is something quite different. Alec talks about the nature of power. Wanda talks about the complexities of planetary health. Erin talks about Indigenous ancestral knowledge, and I talk about often difficult to define interdependencies existing in human systems. None of these things can be measured. None of these things can be understood through formulae or simple cause and effect explanation. None of these things can be represented by elegant models that purport to explain and address the complexities of human behavior.

But the belief that they can is ubiquitous, and potentially dangerous.

Of the three perspectives of the Arctic tundra I mentioned above: through the eyes of an artist, an economist, or a healthcare professional, and there are countless others, two seem to receive the most attention in the media, in our schools and colleges, and in our conversations at the pub or the dinner table. These are the ones that can be represented numerically. The Canadian Arctic encompasses almost 4.5 million square kilometers, includes more than 36,000 islands, and it contains nearly 40% of the country's landmass. These facts, represented by numbers, while certainly important in many ways, do not portray the reality of the tundra, the breathtaking beauty of the northern lights, nor the desolate darkness of the long Arctic winter's nights.

There are some 70,000 Inuit people living in the land we now call Canada, most of whom live in Canada's far north. Life expectancy among the Inuit people is 12% lower than that of Canadians living farther south, likely the result of the availability and cost of healthy foods. More than 40,000 Inuit are comfortable speaking an Inuit language or dialect. While fruit and vegetables and what might be found in grocery stores farther south are expensive in Nunavut, housing remains limited and relatively expensive. As many as 68% of Inuk adults commonly eat what they would call country food; that is seal, caribou, narwhal, berries, and other more traditional foods of the area. Again, while these figures are in some ways revealing, they say little or nothing about the welcoming beauty of the Inuit culture, or the soft-spoken demeanor of so many of the Inuit people.

In these pages, Dr Balasescu explores in detail the complexities of how we came to revere data and formulae to such a degree, the implications this reverence has on how we live our lives, and the impact it has had on cultures around the world. From this understanding, some stark realities that lie beneath artificial intelligence, how it works, and what it actually does, become apparent. Such understanding leaves us wondering what is true, and what is not.

Where is truth? Is that a question to ask, or even one that can be answered?

As an academic, perhaps not dissimilar to other academics, I use the word, "truth" with some trepidation. When we talk about truth, one has to ask, whose truth, and what exactly do we mean when we say "truth?" But here I want to make a point about truth. It seems important in the context of Dr Balasescu's work. Facts on their own are not truth. They are facts. When brought together, some might call them data. But again, points of data brought together are not truth; they are simply points of data brought together. Truth is something more elusive, or perhaps illusive. Truth is an alchemy of the human condition. It is how we think about things that give them meaning, and thereby truth.

And yet, as Dr Balasescu points out, we may be ceding our human awareness of truth to what seems a simplistic reliance on data and facts. Counting implies reality. If something is important or relevant, it counts; if not, it doesn't count. We talk about SMART goals, and they seem to "discount" the humanness of our dreams and aspirations. Some so-called leadership experts are fond of saying, "If you can't measure it, you can't manage it." Or, "If you can't measure it, you can't improve it." I have even heard, "If you can't measure it, it doesn't exist." Facts and data are vanquishing truth in our modern, mainstream, Western society.

As I go about my work here in Iqauit, another drama is playing out far to the south, and indeed around the world. In this drama, we see a post-truth paradigm exploited by a diversity of authoritarian regimes taking hold in different countries. In this emerging paradigm, the facts and data to be rendered by that alchemical process, in many cases are simply not correct, leaving the resultant truth wanting. In other words, they are lying, a uniquely human activity. Artificial intelligence lies too. It learned that dark art from us.

Across the millennia, humans, just like all other species, have evolved. Through a process of natural selection, we have adapted to the environments in which we live. Though I am not an expert in this field, I suspect lying is either unique to our species or more prevalent among humans than it is among other species. But if we think about the emergence of artificial intelligence, the focus of the pages of this book, it is only a reflection, and a weak one at that, of our current stage of evolution. If truth is an alchemy of the human condition, then so too is lying.

Reality is said to be the sum of all that exists in the universe, but not of those things that do not exist. But what do we mean by "exist?" I read recently that scientists have calculated the total amount of matter and energy in the universe. Matter and energy. Is this all of reality, or is there something more? Does meaning exist in a measurable universe? And what of truth? It holds no potential energy, displays no kinetic energy, and has no mass. So positivist science might say that it does not exist.

Similarly, the meaning we find in art cannot be measured. I have a painting hanging in the living room of my modest home, and when I look at it, I feel. That feeling cannot be measured, cannot be commodified, cannot be reduced to data. Nor can that feeling be "accurately" described, because the next time I look at it, I might feel differently. And, of course, when my partner looks at it, her feelings are most certainly different than my own. Such feelings defy measurement, but does that mean they aren't real, they aren't a part of the sum of all that exists? Looking deeply at reality is what Dr Balasescu does on these pages, and looking so deeply creates deep understanding.

This understanding is at the heart of the important messages this book is sharing. As artificial intelligence rises into prominence, this ceding of truth to facts and data is revealed, but this way of thinking is not new. Perhaps it emerged in the enlightenment era, or perhaps earlier. I'll leave it to Alec to explore the origins, but for now, I will simply say that positivist science sees reality as certain, and that certainty very often appears to us as truth.

Years ago, as a younger man, I found myself immersed in my doctoral studies, and at one point railing against all notions of positivist science, against positivism altogether. Surely this dated paradigm had no place in the social sciences. Positivism had had its day and now we were moving on. After all, one simply cannot isolate variables in human systems. And yet as I strove toward understanding, I found that much of what I held to be "true" was grounded in positivist science. My father had been a physicist, a practitioner of positivist science. My mother had been a mathematician, a practitioner of the most positivist of the positivist sciences. Their work mattered. Their work had meaning during their careers, and still did. Why, my father had even been credited by a British admiral as having won the Battle of Britain because of the research he had done developing magnetrons, which had aided the allies in that battle by making radar a possibility.

Positivist science has revealed countless facts, has made possible countless innovations, and indeed has made our lives what they are today. And yet facts are still not truth. I realize that in academic fora, especially those in which my parents existed, I am on very unstable ground. But I think we humans might well consider the nature of knowledge and that alchemy of the human condition that gives it meaning as artificial intelligence rises into prominence. The philosophical studies of ontology, epistemology, and axiology seem relevant, but these pages do not contain philosophy lessons. Rather, they explore patterns of thought that influence our perceptions of reality and that created artificial intelligence, and significant insight into how we might augment those patterns of thought toward now living with artificial intelligence.

Perhaps in the end, the nature of truth is so uniquely human that it will continue to defy all attempts at definition. Perhaps it will be like that quote so often attributed to Louis Armstrong. When asked by a reporter what jazz music was, he responded, "Lady if you have to ask, you'll never know."

Artificial intelligence has now been created. We have created a monster, or perhaps the divine. Or perhaps it is just another tool, one of countless technological imaginings, no different than a PlayStation or a bevel gear. Whatever it is, now that it "exists" is part of our "reality," we might want to consider how we might live with it. How might artificial intelligence fit with that alchemy of the human condition that cannot be measured?

As I near the end of the process of writing the foreword for Dr Balasescu's detailed and important work, I still find myself looking out over the seemingly desolate Arctic tundra, and thinking about thinking. I suppose artificial intelligence could have written the words on these opening pages. Drawing from data found throughout the Internet, it might even have written something more comprehensive, more understandable, more eloquent. But it would not be sitting in a darkened hotel room looking, imagining, dreaming, and struggling to stay awake. Such feelings cannot be rendered from data found throughout the Internet. Rather, they are uniquely human, and in this moment, I find myself truly understanding that while it cannot be measured or rendered into data, humanness has real value.

Royal Roads University Guy Nasmyth
Victoria, BC, Canada

Gratitude

This book emerges from a web of collective efforts that sustained, nourished, and ultimately made it possible for me to dedicate time and resources to its writing. From its inception to its current form, I met different people who contributed to the thinking and doing that went into this text. I will list most of them, but there are definitely more, in the entanglement and webs of relationships that are contained in its making. First, I would like to thank Çise, my companion during most of the years of conceiving this book, for her patience and understanding; the costs were high, and I hope it is worth it.

Roy Bendor sparked my interest in this topic, and Thomas Heinrich Musiolik encouraged me greatly, perhaps without knowing, to keep my focus and to write. Kenza Ait Si Abbou was an inspiration from the moment I contacted her in my intention to interview her for the book. She ended up interviewing me, and teaching me a lot about the current state in AI. Speaking of people I interviewed for the book, my gratitude goes to all of them for the gift of time they gave me, particularly to Dagmar Monett, Andrew Zolli, Adela Fofiu, Denis Rothman, and Djamila Amimer for their patience in the repeated meetings we had. There are more, and traces of our interactions appear in a way or another in this book.

The entire teaching collective at Royal Roads University, Victoria, has a special place here: Wanda Krause's and Erin Dixon's voices are vividly present in between the lines of this book and so is Guy Nasmyth's. Thank you for your friendship, support, and trust! Also, Guy, thank you for your careful commenting on the text, editing for English, and accepting to write the foreword! Lisa Çorak, your organizing mind prevented me from scrambling in my back and forth between continents. Thank you all!

Jimena Chalchi Garcia, thank you for your guidance into ways of thinking and perceiving that go beyond our slice of time and link us to our ancestry and possible futures. Gayathri Gopan, thank you for sharing the wonderful story of the practices around Vilakku in Kerala, and for allowing me to share it along with your family photos.

An important moment of this process was my impromptu visit at Mahidol University, Bangkok, in 2023. Gratitude to my colleague and dare-say friend Eugen

Ciurtin who made this encounter possible. The welcoming from and conversations with Pagorn Singsuriya, Samantha Rajapaksha, Piyanat Prathomwong, Pattira Thaithosaeng, Worapong Charoenwong, and Claudio Cicuzza were overwhelming and inspiring. Their influence on the thinking and writing of this book is undeniable.

I received a lot of emotional and material support under different forms during this time. Susan Ossman was always present as a friend, and her encouragement and example are crucial to me. Iulia Gorzo was constantly pushing me to keep going, particularly in the last stages of the writing. Her work ethic is a deeply motivating example. Adina Greşanu inspired and infused doses of trust in my writing. Anca Bejan gave a close to final form with her careful help with references, comments, and notes, while Iulia Hau provided shelter and support during the difficult times of pitching the book. Speaking of shelter, I have always found sanctuary at the St Antuan Franciscan convent in Istanbul, where conversations on theology and human nature accompany the late nights with the brothers. During one of these, I found clarity on the question of the snake symbol in the Bible—important at some point in this book. Thank you, Fra Iulian, for making St Antuan my home in Istanbul.

I could have not finished the writing in time without the graceful offering of housing from Horică Mixich, who allowed his parents Olivia and Vlad to let me stay in their momentarily empty home during the last 3 months of writing. Perfectly placed far enough from the tumultuous city, it gave me the peace of mind needed. Thank you, Horică. Your toys are wonderful.

Florin Furdui, friend and mentor into the intricacies of the business world has a central place in the making of this book, and our joint venture is discussed in detail in Chap. 8. Thank you for your patience, advice, and presence.

George-Paul Meiu, thank you for the interest you shown in my work, for the comments you provided that guided my finalizing the thinking around matters of subjectivation, and for your invitation to talk about this book in both the colloquium *Anthropological Crossroads* series, and to record an interview for a filmed podcast on the *Ethnographic Imagination Basel* (EIB), both at the University of Basel. I am looking forward to our future conversations.

There are many more people who are present in the spirit of this book, not least my colleagues in healthcare from Vancouver in the team lead by Beth Snow, and the doctors and nurses who generously allow me to shadow them during their stressful program.

I would not finish this list without mentioning two persons whose influence on my work is ever present: the late Vintilă Mihăilescu, who introduced me to the discipline of anthropology at the University of Bucharest and taught me the fine ropes of fieldwork, and Bill Maurer, my Ph.D. advisor from UC Irvine, who taught me how to put together sparks of imagination in coherent theoretical frames. Also, he told me to read noir novels while writing—and this definitely helps in organizing the narration. Although more than three decades have passed since I started my journey in anthropology, your voices are ever present when I “make” anthropology. My gratitude goes to you both.

Introduction

[Hinton] sighed. "We can't be in denial," he said. "We have to be real. We need to think: How do we make it not as awful for humanity as it might be?" in Rothman (2023)

Geoffrey Hinton is one of the "parents" of deep learning, and it is perhaps apt to start any exploration with various origin stories. Everything has a genealogy, and this book started to germinate some time ago, perhaps without me being fully aware of the form it would take. But since I am signing it, I feel the need to set out in the introduction not only what I intended to talk about but also "where I am talking *from*." In academic words, that would be my "subject position."

Back in 2014, I was living in Vancouver and thinking about what to do with my life. At that time, my formal training in anthropology had already finished about ten years prior with a PhD focusing on fashion, technologies of the body, and political subjectivation, and my practice was more at the intersection of art and teaching, while experimenting with various forms of knowledge production and use. I had accumulated work experience in a corporate environment for a car company (I've always hated cars), in state administration and public diplomacy, and in various universities from Paris to Bahrain and Istanbul, as well as in my native country of Romania. I was wandering, perhaps lost. Yet my preoccupations were steady and enduring technologies, art, and the body. To this, becoming a parent added a heightened concern about climate change and its potential catastrophic effects. Around the time my daughter was born in 2015, I met Roy Bendor, who lived in Vancouver then and now teaches at the Delft University of Technology in Holland. We spent many hours walking around our neighborhood in downtown West End Vancouver, where the asphalt meets the dark soil of Stanley Park and the sands of the English Bay beach—names that evoke the Canadian colonial past and present.

Our steps were animated by conversations on current affairs, art, and the future of our world. Roy was pursuing a postdoc at the University of British Columbia studying digitized futures. A new concept on my radar, I was captivated by the possibilities it represented for anthropological inquiry, with obvious links (to me at least) to my past explorations on the body and material culture. To my mind, digitization **was** material culture. Exploring its metaphors, I discovered that this is not necessarily a shared understanding, let alone a truth. But let us not anticipate. The concept of machine anthropology started budding in our minds, as a means to explore the complex relationship that humans have with their own perception of self and environment. Fueled by our imagination and stimulated by a "call for fictional research" for a now-defunct project (its webpage still announces as "coming soon"), we imagined a future in which symbiosis with machines was the answer to climate change challenges. It is only by modifying our behavior that we can mitigate and adapt to climate change, we thought. And so, we imagined the Universal Sustainability Sensorial Regulation system that would solve all our problems. That text got written and still survives on Roy Bendor's blog. Explanation below, irony intended:

> The Universal Sustainability Sensorial Regulation system (or USSR) was developed in response to stalled negotiations for a global treaty to curb GHG emissions, and to the catastrophic failure of geoengineering. Building on recent advances in nanotechnology and the consolidation of the Internet of Things, USSR is designed to regulate the human sensorium based on real-time input from atmospheric and machinic agents. Following an expedited process of ethical approval, the system has now been operational for 20 months, in 5 sites on 3 continents. Our initial findings show that as hypothesized, (1) manipulating human thermal comfort levels by only 2°C saved an average of 1500 kW/h and reduced GHG emissions by an average of 5 tonnes annually, per household; (2) initiating higher levels of gluten intolerance in response to global grain shortages proved successful in preventing food riots; (3) while our comprehensive surveys and interviews indicate that human subjects find the system to be relatively satisfactory overall (m=6.4), we are still grappling with what can only be described as overzealous behaviour by machinic agents (OBaMA), resulting in 31 confirmed injuries. We intend to investigate this further by deploying the innovative techniques of *machine anthropology*." (Bendor, 2015)

I went on to write an article on machine anthropology in which I used the metaphor to describe what would happen if artificial intelligence started generating its own knowledge about us humans (Bălăşescu, 2020), in other words, if they developed their own field of anthropological knowledge. Of course, the metaphor is highly anthropomorphized, but it gave me the space to think about many ways in which the use of technology recreates our perception of self, the other, and the surrounding world: first as imaginary projection and then as discourse and finally as reinforced reality. In many ways, this book is a prorogation of my article that appeared in the *Journal of Future Robot Life*. In as many ways, it is not. I do not believe that automation systems we unfortunately call artificial intelligence (AI) are in any way intelligent, let alone conscious, not by themselves. But I do believe that our intelligence, emotions, and consciousness are constantly reshaped by our relationship with technologies. In Jean-Pierre Warnier's words, we think with our fingers (Warnier, 1999). Even if the fingers only swipe on a screen, the thinking takes place in the interaction, and we constantly transform as a species via our interactions with our material culture, digital technologies included. I prefer to use the term "transform" to that of "evolve" for reasons that I hope will become transparent in the course of the book. In this perspective, our intelligence could be best described as the entanglement with the surrounding world and what we make of it—both figuratively and literally. But, for the moment, in human imaginary the brain remains the center of consciousness, the place that orders the experience, and orders us around.

In the past decade, much has changed in our world, including in the realm of artificial intelligence, and in the conversations about it. The very question of the possibility of consciousness of the automated systems gained primacy and, rightfully so, the governance with and of AI moved to the center of political preoccupations. At the same time, the depiction of AI seems to oscillate between universal goodness and pure evil.

In fact, two apparently opposite tendencies are shaping the world today: the universal appeal of technology perceived as both the ultimate savior and the defining existential threat for humans and the centrality of a return to nature in order to face the challenges of climate change. In the next couple of hundred pages, I will propose an anthropologically informed perspective on the points of convergence of these tendencies in order to examine the complex dynamic at the intersection of nature, technology, and culture. This is not a book about artificial intelligence; instead, this is a book about the tensions and contradictions inherent in the way we define nature and how they are reshaped in the discourse and practices surrounding technology within today's capitalist paradigm. AI is used here as an "object for thinking" about us as humans and the way we conceive ourselves in the world.

More particularly, I will attempt to dismantle the assumptions with which we approach the use of AI. I critique the managerial system woven through the dominant view of nature circumscribed to the realm of a manageable pool of resources, and the effects that AI application may have in continuing to create and reinforce marginal naturalized categories for further management, control, and optimization.

Historically, designing technologies to respond to challenges we have faced has effectively redesigned us. The scope of technology influences the very way in which we design it and the way in which we define ourselves. This book looks closely at the politics of optimization within and outside managerial perspectives and argues that one of the main sources of our repeated failures related to governance and climate change lies not intrinsically in the tools we use but in the scope we design them for. That is, the mental tools used for building our physical tools are to be reexamined, critiqued, and reassessed in order to move away from the real danger of perpetuating the status quo of capitalist extractivism.

We tend to create a mystical aura around technology, abstracting humanity. It seems that now, more than ever, with the promise of the automating technology called artificial intelligence (AI), we are ready to imagine a world that transcends, for good or bad, our humanity. This image of technology reminds us of the magical thinking surrounding natural phenomena—except that, this time around, the object of awe is human made. It is as if we are slowly stepping into technological animism. At the same time, climate change—a somewhat clearer existential threat—is approached mainly as a "problem to be fixed" with "targets to be achieved" using a managerial approach with capitalist undertones well entrenched into a colonial, extractivist view of the world. This indicates a certain cultural bias toward the "shape of the environment" and also makes it tempting to understand and apply a technological fix to it. More often than not, AI comes to mind. When AI is used as a tool to understand and mitigate climate change, we need to first understand the manner in which we conceptualize "nature" and what are the effects of this perspective on the way in which we apply AI models to the task at hand. When AI is put forward as a solution for creating a harmonious and peaceful society, with generalized wealth and zero risk, we are not only one small step away from a future dystopia but also actively turning the corner toward a total surveillance state. It is as if AI promises an "environmentally friendly surveillance capitalism." Thus, how are we to approach the use and abuse of AI in a contextual manner? How can we fine-tune our approach to this very powerful tool?

Here I will propose a perspective that is informed by personal experience in the field of deploying AI solutions, by interviews with people directly involved with AI tools, and by the wealth of literature produced on this theme. The current conversation is rife with moral and ethical concerns that are often too abstract. The generalization of AI use and climate change are two facets of the same challenge, and what unifies them is the less commonly analyzed culturally variable definition of "humanity." This book's ambition is to bring these two apparently separate domains together and to analyze the inherent pitfalls of monolithic understandings guided by a managerial view of nature, technology, and culture.

By virtue of my training, I tend to attribute everything to culture, in the sense that nothing produced or shaped by humans, science and nature included, falls outside of it. Culture in this sense is the mediator of the ways in which we both produce and experience the world. In its endless variation, culture accounts for the diversity of worldviews, ways of making, and ways of being. Culture is unbound, forever shifting and endlessly transforming. It takes plural when we try to describe a plurality of views, and it is better used as singular when we approach it as a framework of understanding. Just like everyone, I have my own worldview, informed by my life trajectory. I grew up in a small town, close to the open hills, enjoying swimming in a river and chasing critters, without any restrictions in my contact with the environment. I also grew up in a totalitarian political system based on surveillance that was dismantled when I was already 15 years old. Perhaps it is not by chance that the themes of nature and surveillance intertwine in the coming pages, in which I allow myself personal musings and contemplations. Sometimes I recount anecdotes and use them as props for reflections—without considering them proofs of concept but ethnographic moments within the thick description of the web of significations, interactions, practices, and metaphors that surround us and form the very thickness of culture. Contextually, culture may be understood as a separate realm from technology or from science for that matter. Particularly if we conceive of *science* as the realm of perfect objectivity and not of sciences—as Bruno Latour brilliantly observes in his *Politics of Nature* (Latour, 2004).

My invitation is to follow seemingly unrelated paths toward the same goal: making sense of what it means to be human in a world that casually blends nature, technology, and biology in an overarching discourse of colonial heritage centering on progress, optimization, and capitalization through resource extraction. Looking at the triad of nature, culture, and technology, this book argues for the harmonization of these elements to provide a stable tripod for the continuation of life on Earth. The triad can just as easily be a recipe for catastrophe: the choice is ours to make.

The text of this book is organized into three acts: "Here and Now," "Past Traces of the Future," and "Where To?," with three chapters each. The text aims to make that choice clearer but not easier (I leave the latter to the machines). The first Act invites the readers to explore three categories that render the ideal AI-driven society problematic: matter, the body, and culture. Each one of them has a dedicated chapter. It is not by chance that, in AI discourses, these categories are either obscured or tentatively standardized. These first chapters also present examples from my work as an applied anthropologist in the healthcare system in Canada and distill my preoccupations with the anthropology of the body. The first chapter, "What's (the) Matter," approaches the intersection of materiality and automated systems with real-life examples, while contrasting discourses about AI and the observations on the field. It brings into conversation the sometimes competing goals of AI promoters and AI adopters, and it shows tensions and adaptations that take place on the ground when an automated system is implemented. It also shows how the goals of automation are shifting, being literally redefined by the introduction and use of AI itself. The chapter goes on and explores theoretically the meaning of this interaction and attempts to stay in the tension created through the renegotiations between what is "designed" and what is "in the field." By reviewing and building on existing literature, the opening chapter indicates the stakes of the "politics of technology" that tends to obscure materiality and points out to a possible critique of large categories such as "modernity," "lab-based reality," "public/private distinction," and ultimately "disembodied divinity."

Chapter 2 brings the question of "matter" at the closest to living beings, thus to humans, by discussing "the body," from both anthropological and technological perspectives. Using both field experience and theories of the body, the chapter discussed conceptions of body and self, proposing a meditation on a fundamental question: "are we a body, or do we have a body?" The answers are obviously context dependent and culturally shaped, but at the same time, the dominant unspoken assumption about this shapes the way in which we culturally approach the materiality of our lives and the way in which we, as humans, are making culture. It also shapes the way in which we understand "nature"; the fundamental argument here is that the human body is the first (and last) point of contact with "nature." This chapter prepares the reader for the more elaborate conversation on finitude, death, and modern capitalism from Chap. 6.

Chapter 3 closes the first act of the book, while opening the possibilities of reading and using AI in a variety of dynamic ways, culturally determined but not culturally bound. In fact, boundaries do not separate cultures; on the contrary, they are the site of cultural innovation. While culture embodies past experiences, it also potentiates future developments and the very shape of the future as the product of imagination. While it presents itself as a critique in which the category of culture is standardized in the corporate language to facilitate AI adoption, it also shows how culturally triggered adaptation reworks the idea of automation, artificial intelligence, and optimization. In other words, this chapter brings back what algorithms obscure or forget altogether: context, not as a standardized set of "cultural particularities" but as a dynamic environment in which all elements involved change as they interact. The success of AI adoption depends on the web of relations between the agents that produce, develop, redistribute, and share knowledge: humans and machines alike.

Act II, Past Traces of the Future, proposes an incursion into the past, particularly the one that shaped the dominant conceptions of nature and of technology. The argument here is that these two cultural and ideatic categories are to be understood together, as feeding one another's borders, limits, and transgressions. This bundle of chapters also argues that the overarching referent of the current construction of these categories is digitization—as in measurement, expressing observations in and through numbers, and mathematical calculus. Chapter 4 deals with the emergence of "nature" as we know it today, dealing heavily with the contribution of Alexander von Humboldt in both the shaping of this notion and in the critique of its current understanding. The tension between what is measurable and what is not and how to account for that which cannot be counted is analyzed here, along with the effects of this tension on the current understanding of the meaning of "nature" and "science." This chapter extends the analysis and proposes an exploration of the effects of digitization of the act of observation on the relationship people develop first with their bodies and implicitly with "nature" and furthermore with "society." Is it a resource? Is it an asset?

Chapter 5 deals with the concoction of "humans," "nature," and "societies," by asking the simple question: When did numbers become explanatory of the world and How "natural laws" emerged? To what effect? This chapter is the turning point of the book, providing an understanding of the process of knowledge production in the current paradigm and its origins in merchant accountancy. Largely discussing the history of post-Renaissance Europe thinking and transformations in relationship to numbers, this chapter is a critique of data-driven societies that are largely built on a model that confounds the exactitude of the mathematical language with the accuracy of observation. This, as the argument goes, was the practical solution to the unsolvable problem of induction. This also obscures the particularities of the individual observations. By proposing "the lab" as a model of the world, the world is remodeled as a lab in the attempt to apply the findings independent of context.

Some of the social, cultural, and political impacts of this approach are touched upon in this chapter. They also offer a segway into Chap. 6 that brings the conversation back into the current state of affairs. It proposes both a review of the ethical matters of "surveillance capitalism" in the light of the argument of this book so far and a meditation upon death as an "uneasy" phenomenon for digitized modernity. Biological bodies that insist on dying remind the digitized system of its imperfections, of the materiality of its ultimate referent, and of its finitude. "Solutions" are always proposed, from bioengineering ourselves out of the dying process, all the way to "salvation" into the machine. In this type of discourse, the quasireligious aspects of AI become transparent. It is at the tension of individual dying that we are to look for the very tension inbuilt into our understanding of technology, machines, nature, and intelligence (be it artificial or otherwise). This analysis prepares the reader for the Act III.

The last three chapters of this book, reunited under the title "Where To?," propose the decentering of the view on AI. This decentering has a series of sources, for which the reader is already prepared by now: the decentering of modern thinking by its historicization, the decentering of the human species from the idea of environment, and the decentering of the human individual itself from the modeling of the world. Chapter 7 deals with the need to decenter Enlightenment and post-Enlightenment ideas from the way we understand ourselves and the world. It proposes a foray into the other than modern, and ultimately other than individual-centered ways of thinking, with an incursion into Indigenous and Local Knowledge systems. The ultimate argument of the chapter is that imagination is freed once we decouple intelligence (and consciousness) from the individual, and once we remind ourselves that personhood is a relational construct, not an ontological given. Examples from around the world, ranging from memories of my great grandparents in Eastern Europe to accounts of students from India, combine with theoretical and practical approaches on other-than-modern and other-than-human ways of thinking and being to form the backbone of this chapter.

Chapter 8 accounts my experience in introducing the human decentered vision in the practical world of energy transition to renewables. This chapter is based on an ongoing effort, doubling as fieldwork, to convince organizations that build and exploit solar farms to create habitat for biodiversity in between the panels. The findings so far are both frustrating and encouraging. While enthusiasm is usually visible, finance matters act like a big force of reversals. However, the process is interesting because it shows a gradual opening in the world of business to ideas akin to considering "nonhuman stakeholders." The future is in the making.

The final chapter of the book proposes a wrapping up of the conversations opened here, the final weaving of all the threads apparently left hanging in the writing. The questions return constantly to the relationships with the tools we make (technology), and with the world we inhabit (nature, environment): Where do we place ourselves? We are at the articulation of these categories, and we are also responsible for articulating their relationship. Do we conceive ourselves as part of this material world and strive to harmonize our products and tools with it? Or are we placing ourselves outside, implicitly building cultural and material oppositions between "technology" and "nature?" This is a hard choice to make. I hope that reading this book will make clearer the consequences of our choices.

References

Bălășescu, A. (2020, May 13). Machine anthropology or will robots talk about us behind our back? *Journal of Future Robot Life, 1*(1), 1-X.

Bendor, R. (2015, January 3). Fictional research. *Digital Sustainability*. https://digitalsustainability.com/2015/01/

Latour, B. (2004). *Politics of nature: How to bring the sciences into democracy* (1st ed.). Harvard University Press.

Rothman, J. (2023, November 20). Why the godfather of A.I. fears what he's built. *The New Yorker*. https://www.newyorker.com/magazine/2023/11/20/geoffrey-hinton-profile-ai

Warnier, J.-P. (1999). *Construire la Culture Materielle. L'Homme qui Pensait avec ses Doigts* [Constructing material culture: The man who thought with his fingers] (French). Presses Universitaires de France.

Competing Interests The author has no competing interests to declare that are relevant to the content of this manuscript.

Note on the Text

Some parts of this book have been published by the author under different and modified forms as follows:

Parts of Chaps. 1 and 3 appear in Balasescu, A (2022) *Ethics, Health, and AI in a COVID-19 World. Why Contextual Dynamics Matter, and Culture Beats Algorithms?* in Ethical Implications of Reshaping Healthcare With Emerging Technologies, Thomas Heinrich Musiolik, Alexiei Dingli (editors), IGI Global, DOI 10.4018/978-1-7998-7888-9, pp. 1–24.

Ideas and parts of Chaps. 5 and 6 are taken and reworked from Balasescu, A (2007), *Paris Chic, Tehran Thrills. Aesthetic Bodies, Political Subjects*, Zeta Books.

No part of this text have been generated using AI capabilities

Contents

About the Author

Alec Balasescu studies human–technology interaction, climate change and health using anthropological and art-based methods. Drawing on his hands-on experience in administration, cultural diplomacy, health care, and the automobile industry as well as academia, he develops a unique understanding of the dynamic way of how contexts for action are generated. Dr. Balasescu is the founding director of Re-Genera Group, and has been an associate faculty member at Royal Roads University, Victoria, BC, since 2018. He previously served as the executive director of the Romanian Cultural Institute in Istanbul, the chief of cabinet of the Ministry of Culture in Romania, the Lead Data Analyst at Providence HealthCare Vancouver, and responsible for Socio-Cultural and Economic Forecast at Renault Technologies, Romania. He has taught at UC Irvine, where he earned his Ph.D. in 2004, as well as Simon Fraser University, Galatasaray University, American University in Paris, and the Royal University for Women in Bahrain.

List of Figures

Part I
Here and Now

Chapter 1
What's (the) Matter

In 2006, I was living in Bahrain, when I met Tameem, a man who would become my good friend. We took a trip to Romania during that summer. To put it in recent history's perspective, that was the very year in which Saddam Hussein was tried and executed and 1 year before Romania joined the EU. It was also the same year when Facebook introduced News Feed in the interface, provoking a series of events that forced the company to consider users' privacy and control.

As we were strolling up and down the streets, his keen eye observed the cables hanging from the city's utility poles. Coiled, meshed together or simply hanging loose, the urban Internet infrastructure in one of the countries already known for its superfast connections in the world and for its hackers was laid in barren view but barely seen by locals, including by me. My ten years of absence from Bucharest did not entirely grant me the fresh eye for every detail. Growing up behind the Iron Curtain, I was used to layers upon layers of improvisation when it came to infrastructure. While the country was indefinitely under construction, worksites were accessible and unguarded (they were also our beyond-dangerous play sites), allowing us to see how electricity, water, and waste management were linked to the new buildings. There was also no effort to hide it. On the contrary, the entire process of production and functioning was displayed, both on the sidewalks and on the state TV channel at the time that celebrated the proletariat, the ideological backbone of the regime.

Fast forward 17 years after the fall of the Berlin Wall, in 2006, my eyes did not register as out of place those newly hanging cables that decorated the skies in Bucharest and that formed one of the fastest optical networks in the world, second only at the time to South Korea. To me, they were normal. They witnessed the change, socioeconomic and political, for better or for worse. They showed how Romania was connecting to the world. To Tameem, they were exotic, vestiges of a not fully modernized space inscribed in the cityscape. Another 17 years later, in 2023, most of the cables in Bucharest disappeared underground. The city looks modern now, and photos of its imposing architecture can be taken without fret that black coils would bomb them. At the same time, most of the workers who laid them migrated to Western Europe and are now gradually replaced by a quasiinvisible workforce arriving from South and Southeast (SE) Asia.

A. Balasescu, *Climate Change in the Age of Artificial Intelligence*, Artificial Intelligence, Simulation and Society,
https://doi.org/10.1007/978-3-031-90042-6_1

Speaking of SE Asia, this chapter is written during a trip to Cambodia and Thailand, thus imbued with reflections generated by my encounter with the urban spaces here, and stimulated by the encounters I had at the department of Religious Studies, Mahidol University, Bangkok. While Thailand is considered now a medium-income country, Cambodia is classified as one of the poorer nations. Both places display aspirations of modernization translated by shiny LED screens, clean-cut advertising billboards, and the adoption of electronic payments. They both are connected to the world's virtual space, and their inhabitants use social media on a regular basis. However, more often than not the giant screens tower over shanty towns scattered in pockets inside the megalopolises, and in both places, the cables are alive and well, that is, they are visible, inserted in the landscapes that sometimes offer suggestive juxtapositions, etc. (Fig. 1.1).

But what does this all mean? Why the drive to obscure materiality? And what does it have to do with AI or climate change for that matter?

The changes brought about by digitization of industrial and human activity, also known as the fourth industrial revolution, have engendered a false sense that our world—at least the part of it that is presented as general aspiration—is virtual. AI amplifies this perception, creating the belief that, after all, everything can become dematerialized sooner or later: from financial markets to life itself. Virtual seems to be the new real, from cocktail parties and sex to education and spirituality. Misleading metaphors such as "the cloud" perpetuate this perception, and more often than not, we neglect the fact that every byte of data is material, needs support to move around, uses space to be stocked, and consumes energy while processing.

In their article *Energy and Policy Considerations for Deep Learning in NLP*, Emma Strubell, Ananya Ganesh, and Andrew McCallum remind the impact that the training of Natural Language Processing models (NLP) has on the environment: a single algorithm training emits 300,000 kg of carbon dioxide equivalent, about five times more than a car in its entire lifecycle (Strubell et al., 2019). This is without counting the tuning and experimentation of the NLP. The authors concentrate on energy consumption and subsequently on the control of computing power, since servers are privately owned, and this generates a type of monopoly with direct implications in research and development dynamics. As with any other monopolies, a "rich get richer" logic emerges here, too. The article argues for a model of open access computing capacity in order to reduce energy consumption and to encourage innovation.

The results of this study are significant and go well beyond its scope. In an article with deep implications on policymaking, Mark Coeckelbergh analyzes the pros and cons for using AI in tackling climate change, starting from the matter of energy consumption (Coeckelbergh, 2020). He argues for the necessity of acknowledging the energy use among the professionals and those who use AI and moreover to introduce its calculus as metrics of success as steps toward an ethical approach on energy consumption: rendering it visible. The author pushes the reflection deeper into the questions of ethics and policies, pointing out the way in which we understand and

Fig. 1.1 Advertising for virtual reality—Siem Reap, Cambodia—photo by the author

use the tool, be it called "intelligent" (AI) is what matters most. The case in point is that regulating human behavior in relation to AI is what will offer (or not) solutions for the complicated matter implying human activity and climate change. The article argues for a fine balance between regulations and freedom, between nudging toward or imposing a certain behavioral change, and between human agency and a possible Green Leviathan (remember the USSR fictional system Roy Bendor and myself imagined…). I will later return to these points and the conversation Coeckelbergh proposes, that of human hyper-agency[1] over the planet, but for the moment I would like to linger on the question of AI materiality—and its invisibilization.

[1] "[…] our modern desire to control over everything and everyone by means of science and technology. [...] Climate change can then be interpreted as the outcome of, perhaps the pinnacle of, this

Energy consumption is only part of the materiality of AI systems. The entire network on which AI is based implies much more than this. Without the material support of servers and cables, workers to build and lay them, technicians to care for them, and tons of water and energy to keep them cool, and without constant reference to this very material world, the virtual world would not exist. Kate Crawford offers a mapping of these systems at a global scale in her book *Atlas of AI: Power, Politics, and the Planetary Costs of Artificial Intelligence* (Crawford, 2021). This network of networks adds substantially to the *anthropomass*, or anthropogenic mass—the total of human-made matter existent on the planet. A research at the Weizman Institute of Science made accessible on their website[2] shows that currently the anthropomass exceeds the biomass present on our planet. The shift in balance happened recently in 2020, and the growth was exponential, the anthropomass doubling in the past 20 years. One of the hugely significant materials is the asphalt that emerged in the 1960s as predominant material for paving the roads. Intuitively, the personalization of computers and later the mobile phones, together with the networks sustaining the digitization of our planet, contributed significantly to the accelerated growth after the 1980s. The impact of the multiple layers of materiality of AI is yet to be uncovered; however, there are both efforts in researching them with anthropological tools and activist reactions that intensified in the past 2 years (2023–2024), the latter probably due to the wide acknowledgment of this impact both on emission and on water consumption (O'Brien, 2024). Currently, there is a National Science Foundation (USA) grant awarded to a team lead by Tamara Kneese from the newly created *Algorithmic Impact Methods Lab*[3] "to holistically measure and understand the social and environmental impacts of Artificial Intelligence (AI) technologies, particularly concerning the massive data centers and other infrastructures that must be built to support them."[4] Until we have the opportunity to see the results, let us turn our attention to the materiality of our world.

A Material World

This is the materiality of the systems. To this, we need to add the materiality of its applications. That means, what happens when the AI systems go out of the lab. How do they interact with the "real world," or in other words, what happens when we throw people and preexisting material structures and designs into the mix?

---

strong grip we have gained on the planet: it is the result and manifestation of this hyper-agency that has put nature and the planet under our full control and that has pervaded the earth and its ecosystems to such an extent that even the climate is now the result of our agency." (Coeckelbergh, 2020: 73).

[2] See AnthropoMass [Website]. (n.d.). Shift in balance. Retrieved December 20, 2023, from https://anthropomass.org/shiftinbalance/

[3] https://datasociety.net/research/algorithmic-impact-methods-lab/#

[4] https://www.nsf.gov/awardsearch/showAward?AWD_ID=2427700—accessed 13 Nov. 2024.

For 2½ years, I worked on the introduction of an electronic record keeping system in more than 40 health units in British Columbia, Canada. I was present from the inception to the implementation in two different units, and able to observe, as part of my job description but also from pure anthropological interest, the changes that occurred on multiple levels, from work relations, including those between health provider and the patients to the relationship with space. In other words, I witnessed the emergence of a new knowledge production system, containing the elements of the old one, but significantly changing its processes, functioning, and thus understanding of the activity. Dematerialization played an important role in the decision to adopt this system. It was one of the strong selling points, bearing the promise of getting rid of the cluttering paper-based records, with all that they implied, from freeing space for movement to avoiding errors linked to indecipherable handwriting—medical doctors scoring high on the scale of ermetic calligraphy. Despite its clear advantages (i.e., instant access and sharing of patient information, decipherability, assisted decision-making), this turned out to be only half of the picture. A different materialization was waiting around the corner.

Currently used all over the world, the electronic healthcare systems combine the patients' records, also known as patient charts, with the automation of certain operations, data extraction, and storage, in order to facilitate and enhance the quality of decision-making based on evidence, subsequent healthcare, and medical research. The system itself embeds a series of decision support features, in the form of alerts—such as alerts in case of the risk of sepsis or for the possible wrong dosage administration. The system user has the option to ignore, override, or follow the instructions attached to any type of message; however, the system will register the option, sometimes with unexpected consequences. (I will return to this in a moment.)

The automated system I worked with entails a specific medication workflow that combines the medication itself with scanning both the medication labels and the patients' wristbands. This requires the manipulation of a computer on wheels with an attached scanner that displays the patient's record, occasional alerts, and standardized warnings based on real-time analysis of the medical data input. Medication itself is part of the real-time data input in a patient's records. The computer on wheels (aka COW) is rather clumsy—with a large display intended to expose the complex patient medical history all at once, not to mention the weight of the COW, so much for AI immateriality.

To "optimize medication," the healthcare professional is meant to follow precisely the workflow briefly described above in order to scan as many instances of medication as possible. The operation itself consists of the following steps in this order: accessing the medication from storage, bringing it together with the COW wherever the patient is, scanning the patient's wristband, scanning the medication, checking the patient's identification, asking verbal identification from the patient (if the patient's state allows for it), and administering the medication as prescribed. Each vial or pill needs to be scanned. If in any of the instances above the machine gives out an alert, such as wrong medication, wrong patient, or wrong dose, the healthcare worker (usually a nurse) has the option to override the system or to insert data manually. Nevertheless, the new standardized workflow emphasizes scanning over manual insertion.

In the view of the Quality Insurance (QI) department, the measure of work efficiency is the rate of medication scanning per distributed dose of medication in each department. Thus, the new workflow is supposed to have a twofold utility: recording real-time accurate data and ensuring the accuracy of records. This is why the QI department considers the manual insertion of the information postmedication a failure in the correct procedure and consequently from healthcare best practices. Moreover, scanning is raised to the rank of efficiency measure and that of a fundamental aspect of work ethics in this environment.

QI narrowly defined efficiency independent of context and of the complexity of the process involved in work. This soon proved to be the Achilles' heel of the system. First, the system was designed in the United States, where medical doctors are not allowed to write prescriptions for medication in range doses. On the contrary, in Canada, variable dose prescriptions are rather common, leaving the decision regarding the appropriate dose to the caregiver based on the patient's condition. Because the system could not be tweaked to accept variable doses, the alerts were set to fire up as "wrong dose" for any medication administered in lower doses than the maximum prescribed, and the nurses were instructed to manually override them.

Therefore, the system generated statistics that showed an increased percentage of "wrong doses," without the ability to differentiate between genuine wrong doses and legitimate overrides. The intersection between design and practice proved to be the generator of an error that fed into judgments regarding the efficiency and the quality of the service of frontline healthcare workers.

Second, in the first couple of weeks, the system-generated statistics showed a drastically decreasing rate of completed medication scanning workflow in the emergency department. The QI department promptly concluded that there was a problem, edging on outright blaming the floor personnel for "not doing their job," as data "showed." Now, data usually points out where there might be a problem, but they do not precisely indicate what the problem is (this is a common source of confusion underlying many decisions taken while using what "data says"). In consultation with the hospital management, I conducted a "rapid ethnography" (Palinkas & Zatzick, 2019) adapted to the context, aiming to understand where the personnel experienced impediments in the medication administration workflow.

The findings underscored the importance of understanding context—including physical context—when we decide to implement any technology and particularly AI-based technology. Although AI systems function in a perceived virtual space, they comprise a great deal of materiality. The COWs, to begin with, seemed, at the first glance, rather sleek, with ergonomic design, variable geometry that could be adapted to sitting or standing positions, nicely rolling wheels, and large screens capable of simultaneously displaying all patient chart information. They may seem close to perfection. And they were, if it was not for the context. In an ideal healthcare site, rendered on a computer, they probably looked like the pinnacle of hospital technology. In the physical reality of this emergency department, designed in the late 1970s, with narrow corridors, even narrower doors, and multifunctional spaces, the COWs appeared clumsy, cumbersome, and somewhat out of place. I noticed that both the computers and their charging station had names, written on stickers and attached to them. Also, the electrical plugs had labels to mark them as charging stations for a specific COW (Fig. 1.2a, b). When asked why, a nurse prompted me to think about what happens in case of a blackout. It would be much easier, she said, to tell your colleagues "Jumbo is working," and eventually rapidly find him, instead of "LC32450X4 is working." Also, knowing that "Henry" the COW is in its place or not may prove crucial in crisis moments—a great example of human-machine interaction effects and the importance of integrating technology in the preexistent cultural forms.

While for industrial designers form still follows function, for new virtualized technology its function must follow cultural forms once it leaves the lab conditions. More on this in the next chapter.

Let us return to, and review, the medication workflow. In the ideal case, it implies that the nurse first checks the patient system using a barcode reader attached to the COWs, conducts the verbal identification by asking the patient their name and date of birth, scans the patient's barcode on the bracelet attached at admission, scans the medication label, and finally administers the medication. After the first medication is administered, the nurse will only repeat the scanning for vials and packages. The entire operation must happen with the patient in sight, at their bedside, which means that the nurse needs to maneuver the COW to wherever the patient is. This workflow may work well when patients are already hospitalized and are waiting patiently to receive their doses. However, this is not always the case.

First of all, some patients arrive in a critical state and all the operations listed above take time. Minutes, even seconds, may be crucial in an emergency situation.

Fig. 1.2 (**a**) "Jumba" the computer in one of the hospital floors. (**b**) The plugging station reserved to Henry, one of the computers on wheels (COW). Photo by the author

Fig. 1.2 (continued)

In many cases, the first life-saving medication must be administered even before the patient is registered in the system that is before having a wristband to scan or an electronic chart for the nurse to refer to.

Second, leaving aside the life-threatening situations, patients arriving at the emergency ward are assigned a priority rank based on a first rapid consultation and assessment. In the hospital I am referring to, a five-tier system is in place, with the first tiers receiving immediate assistance, while the last three need to wait longer. The procedure of assigning ranks is called triage. Due to lack of space, the waiting room for the patients already triaged is the same with those who have just entered the department. The place can get really crowded at times, and maneuvering the COW between the waiting chairs rows proves sometimes unnerving. Medication is administered either by using manual input or by altogether overriding the system.

Third, there are cases in which a patient's condition may rapidly deteriorate. An interesting design feature comes into play here. The COWs work with a USB antenna, and the antenna is inserted by design into the side of the large screen. Many times, while running along the corridors and entering the narrow doors to reach the patient, the antenna would hit the threshold and break off inside. There is obviously no time to search for another available COW.

Fourth, some barcodes on vials and packages may not work. It is frustrating, particularly in emergency situations. Also, due to the design factor explained above, range doses and constantly popping alerts distract the personnel and need to be overridden regularly. These instances are recorded by the system as "wrong dose" or "override" and may be interpreted by the QI department in many different ways.

This machine-mediated relationship between humans is further complicated by newly standardized workflows. Sometimes, healthcare personnel identified workarounds that would allow them to feed the machine with required data, while responding to their own definition of efficiency informed directly by the work floor experience. This happened beyond the emergency department floor. In order to avoid the cumbersome and often frankly impossible maneuvering of the COWs in patients' narrow rooms, nurses found an alternative solution for the medication administration. They would bring the medication and the COW to the entry of the room (usually populated by one or two patients), print duplicates of the wristbands, scan the duplicate and each medication outside the salon, and then take them to the patient inside, without bringing in the COW. They would come out and repeat this procedure for each subsequent patient. This procedure would not meet the standards of the workflow designed for the use of the new system, but it would nonetheless register in the system as the correct procedure.

As a result of the rapid ethnography in the emergency department, an agreement has been reached: unless major modifications in the system were made, medication in the emergency department would not necessarily need to follow the scanning workflow. In order to reach a happy ending like this one, the stakeholders involved must agree/be open enough to reconsider the meaning of optimization and adapt it to the context. Unfortunately, this is not the case everywhere, every time. Also it certainly did not address all other instances of workarounds or the series of unintended consequences that the standardization of the procedure led by the automatic system adoption had on the healthcare processes.

Systems and the City

One of the important aspects of any systemic analysis is deciding on the limits of the system. In the case of the automatic system discussed, does the system comprise the entire hospital or only the department in focus? Does it include the pharmacy that prepares the daily medication according to the system's records? Does it include the relationship with other hospitals and care centers that adopt the same interlinked system? What about the urban setting? Should it also be included? The answer to all of the above is, well, yes. In fact, the best approach is in terms of *nested systems,* in which each system is part of a bigger one, to identify their intricate web of relations.

The automated system promises a "closed medication loop." This means that the medication prescribed by the medical doctor is introduced into the electronic system that can be accessed instantly by the hospital pharmacy where medication is verified, prepared, and readied for delivery daily. Any modifications to an individual's patient doses or medications are registered automatically in the system and updated daily. This saves a lot of work for the night shift nurses. Prior to the electronic system, they had to manually insert the new subscriptions in special folders or check if there are modifications for already registered patients and then send these folders to the pharmacy, every night at 3 am. Based on these folders, the pharmacy technicians prepared the required medications everyday. In the particular hospital, I worked with, prior to the introduction of the automated system, the cutoff hour for the morning medications to be ready was set at 7:30 am. This meant that the technicians would start their working day at 6:30 am.

The introduction of the automated system imposed a region-wide standardization: it required the medications to be ready at 7:00 am, instead of 7:30 am, in order to harmonize the hour of the first medication for all the patients included in the new system. As a consequence, the technicians in charge of preparing the medication needed to start their working day at 6:00 am instead of 6:30 am.

Let us take a look at the bigger system that nests the hospital: the city. This particular hospital is located in a prime real estate region, where houses and rental prices are prohibitive for many people, including our nurses and technicians. Most of them commute to work using public transport, which is a fairly good option and, pricewise, a much better alternative to owning a car. The commute is long, and the first buses only arrive in the vicinity of the hospital at 6:20 am giving no option for the technicians to show up before 6:30 am. Hence the unexpected happened: all technicians from the hospital's pharmacy resigned. The hospital's quick and the only possible fix was to hire pharmacists at the beginning of their career—who most likely owned cars or could afford one—and offer them entry salaries that were automatically higher than those of experienced technicians. This, in turn, created a certain professional tension and disillusionment among the new hires: although pharmacists, they were required to perform technicians' duties and, despite their limited experience, the frustration was understandable. Where would the experience come from, if not nurtured from the beginning of their career?

In this example, we can count at least six nested systems: the medication closed loop nested in the electronic health records system, nested in the hospital system, nested in the network of hospitals adopting the same electronic health records system, nested in the urban setting with its real estate and transport peculiarities, and nested in the system of labor and social inequalities reflecting back on the entire ensemble. Each modification in one of the systems reverberates with unforeseen consequences at every level of the others.

Rematerialization

The dematerialization promised by digitization, like all stories conceived to sell a product or an idea, is only half of the story. In the previous examples, the electronic system settings and automated responses intersected with workflow designs and interior layout and organization of the emergency department. In this case, the standardization of procedures meant to adapt to the need for consistency in the networked medical system reverberated with urban settings and labor conditions and created a perfect storm that destabilized the financial and HR managerial resources of the hospital in unforeseen ways.

Both examples reveal the importance of understanding automated systems, AI included, in their entirety, that is including the materiality that accompanies them. It is true that the circular folder shelves that housed the paper-based patient charts disappeared and with them the fax-machines (yes!) still in use for interdepartmental communications, the breaks in the flow caused by the necessity to ask a colleague "do you understand what is written here? What symptoms? What medicine to give? What frequency?," and the delays in accessing any information about the patient. What emerged is the material support of the AI systems in its interaction with the larger environment.

The system did not dematerialize the process but eliminated a dimension of it, while rendering relevant a whole new material reality: while everything linked to the paper-based world vanished into the virtual space, the physicality of the hospital space, from its interior architecture to its urban settings itself kicked in with a vengeance. Looking only at the lights on the screens (as the QI department initially did), in other words only at the data, largely obscures the materiality, its role, and its often unintended consequences. It also responds to a desire of portraying the world as perfectly formed, cleaned of the material residues. However, this materiality has an important impact on the functioning of AI systems. It is present in all its lifecycle, from its inception, through the training, testing, implementation, adoption, adaptation, and usage. The question is how does the desire for immaterial emerge? Where does it come from? What are its meanings?

The Lotus in the Mud

At this point, I would like to turn our attention to the lotus flower. We find it as a symbol of purity and creation in Hinduism, and Buddhism, on the walls of Zoroastrian Persepolis, or in the Pharaonic art of Egypt, but also as decorative element in the Islamic art, all the way into Europe, surviving on nineteenth-century buildings, including in Bucharest (Fig. 1.3).

It does not bear the same significance in all the contexts in which we find it, in many cases its decorative value overriding the symbolic one. The journey of the lotus flower from highly symbolic to highly decorative is synthesized in the article From Divinity to Decoration: The Journey of Lotus Symbol in the Art of Subcontinent by Aamir and Malik (2017). At a first glance, the perfect form of the lotus flower can be held responsible for its signification. With its many petals arranged with mathematical precision, facing the sky at the end of a long stem that makes it float above the water, immaculate, the lotus appears as pure form and color. However, the water below is, more often than not, muddy, even pestilential (Fig. 1.4).

> All the nymphaeas live in still or slow water and have a condensed shoot apical meristem underwater close to the substrate. The stems, petioles, and leaves are characterized by wide and efficient tracts of aerenchyma and the leaves, and usually the flowers, float on the water

Fig. 1.3 The Lotus on the ceiling of the Romanian Aetheneum, quarded by the "dragons of culture." More about dragons later. Photo by the author

> surface due to their high intercellular air content. Their 'homobaric', i.e. widely interconnected, intercellular system enables them to send oxygen-enriched air by 'pressurized ventilation' from the younger leaves to the subaquatic parts which live in an almost anaerobic stagnant environment (Kandeler & Ullrich, 2009: 2461)

More than the perfection of the flower itself, the contrast with the still waters from which it emerges, with its "anaerobic stagnant environment" below, renders the process of creation visible, and the aspiration to purity tangible. Creation is a process of distillation, and the purity of the lotus flower is its aspiration. Hinduism and Buddhism emphasize both the process and the result, while acknowledging the imperfection of our world and projecting perfection in the divine realm, usually attainable through the burdensome process of purification within the profane world: repeated reincarnations, asceticism, and meditation; they are all ways in which the nutrients of "the underwater" are purified toward the aspiration of perfection. This structure of thinking about the sacred and profane has its correspondent in Euro-Mediterranean monotheisms, the Abrahamic religions: Judaism, Christianity, and Islam, at least up to a certain historical moment. It has its expression in the various ways that living spaces are structured. On the one hand, we have the public/private realms that largely correspond to the sacred and, respectively, profane realms: the sacred space of home, the profane of the public.

On the other hand, we have a clear architectural and urbanistic expression. Contemporary travelers from Europe are constantly shocked, amazed, or amused (depending on whom you are talking to) by the apparent disorder of cities in Eastern

Fig. 1.4 Lotus in Mahidol University Campus, Bangkok—photo by the author

or Southern regions of Mediterranea and beyond. However, if chance or curiosity pushes anyone to enter a space of prayer, a home, or more so a cemetery, the contrast with the brouhaha of the city is all the more impressive. The sacred spaces are perfectly ordered, serene, and clean, the lotus flowers anchored in, and emerging from the mud of the city.

An even more explicit image is that of the city of Phnom Penh, viewed from the bridges crossing the mighty Mekong river. On the shores, close to the waters, the shanty towns drowned in poverty, emerging from their exploitation, in the background we gradually see the "clean" modernized and aspirational city (Fig. 1.5).

Many years ago I visited Behesht-e Zahra (Zahra's Paradise), the cemetery of the martyrs in the Southern part of the city. A young friend invited me there. She wanted to show me another facet of her city. The urban chaos of Tehran stops at the entrance. Beyond the gates of the cemetery, one steps into a perfectly ordered environment with numbered alleys arranged in a grid pattern, a Manhattan of the dead on a surface of 580 ha (one-tenth of Manhattan Island). While strolling among the monuments and tombs of the victims of the tragic Iran-Iraq war, she rhetorically asked: "See how well it is done, how ordered, how *modern (*my emphasis)? Why can't we build the city in the same way?"

Perhaps, for our purposes, it is best to turn her question on its head: "Why would we build the cities as an expression of purity? Why do we "clean" our surroundings of the processual mud? Why is modernity equated with a purified image of the creation, without its process? And to what effects?"

Fig. 1.5 Phnom Penh—view from a bridge—photo by the author

Starting with the interwar modernism, and moving to the acceleration and generalization of its aesthetic ideals in the postwar period—smooth, fast, and clean—technological modernization is characterized by, and sells as, mastery over matter by first obscuring its materiality and then by rendering it irrelevant. Acceleration is the expression of this victory; instantaneity is the ultimate objective. The "speed of thought" is the unattainable aspiration for moving matter around, the stuff of aspirations in fantasies of "beaming" objects and people around, and the object of awe and fret in UFO sightings. "Mind over matter" is the mantra of limitless progress, anything less than this is considered failure… I cannot help but recall the link between speed, Italian modernism of the 1920s and 1930s, and fascism.

Every jump in the technology is accompanied by another layer of obscuring materiality, in both design and user experience. Elimination of sound, elimination of effort, and reduction of the direct manual interaction with the tool, all the way to completely disengaging the body. This is the story that modernization of objects is telling us, from typewriter to computer, to touchscreen, to voice command, and to neuralink—moving the tools solely with the power of the mind. The modernization of cars meant the hiding of the engine, the automatization of the gears, and the elimination of manual interaction—including the elimination of the contact key. The modernization of computers meant not only the move from keyboard to touchscreen but also the hiding of the code behind the interface. This contributed significantly to the accessibility and democratization of the use of these tools. It also contributed to the disappearance from the collective imagination of the interlinks between the result, the product, the perfection of creation, and its process. It has obscured the mud that feeds the lotus. But, out of sight, out of mind…

This ideal has deeper historical roots, tracking back in Western philosophy to the period of Enlightenment, and in religious history at the Protestant Reformation followed by the subsequent secularization. In "The Human Condition," Hannah Arendt (2018 [1958]) dedicates a significant part of her analysis to the impact of technological thinking on the way the West conceives of (human) life and politics. She identifies three major events that mark the entry into modernity: the invention of the telescope, the conquest of the American continents, and the Protestant Reformation.

Each of these occurrences made possible major transformations in sciences and worldviews: the telescope decentered Earth, proved the practical possibility of placing the Archimedean point of leverage outside our planet and thus made possible the idea of objectivity in sciences, the conquest of America began the shrinking of our planet and the possibility of its containment, while the Reform cutoff salvation from the experience of living, since a soul is already saved (or not) at birth. Their combined effect led to what Arendt calls "world alienation." Not the Marxist alienation from one's self, an erroneous way to understand modernity Arendt argues, but on the contrary, the obsessive attachment to the self-decoupled from the surrounding world. The centrality of the self gains its ultimate expression in Descartes' philosophy that moves the proof of existence from its material expression to the expression of the thinking mind.

The combination of the Reformation and Cartesian philosophy paved the way for secularization but in a different, perhaps opposite, direction than that scholars are used to thinking. Or than that the politics of nation-states that embrace this principle is used to formalize it. Generally, secularization is understood as the complete elimination of the sacred realm from our daily lives. But what if in fact what happened is the complete opposite? What if the aspirations of modernity are in fact aspirations of complete sacralization of the world, of transforming our surroundings into the ideal image of purity of creation, in a generalized Lotus absent the mud?

This is a hypothesis entertained in a different form by Meghan O'Gieblyn (2021) in her book *God, Human, Animal, Machine*. That (the Puritan) God is not dead, but found its expression in the move to digitize every aspect of the world, that is to purify it. The author traces this idea that mathematics could purify, and thus sacralize, human existence back to Leibniz and his proposition to assign numbers to every concept, in order to eliminate any possible source of error. I will get back to this idea in Chap. 5.

I would push the argument even further: thinking about the process of modernization and secularization as a process of sacralization of the world in puritan terms would allow us to better understand current cultural, social, and political tendencies. The obscuring of materiality and processuality happened in all aspects of life as aspiration to recreate (a certain imagery of) paradise on Earth: delocalization of production, gentrification of cities (a sort of small delocalization of production), the fetishism of smooth surfaces from objects to buildings in order to eliminate friction, the divine invisible hand of the free-market economy, and ultimately the erasure of public and private are all forms of sacralization.

This form of reading of reality allows us to see that the latter did not happen as the publicization of private life but as the privatization of all that was public. The rules of the sacredness of the private space extended to the public and not the other way around. It is the private that invaded and eliminated the public, turning every aspect of the public into a private matter, and banning from the public the possibility of friction. One can only exist in public as a private persona and not the other way around. Identity politics itself reflects this tendency in which the sacredness of the self (remember the importance of the self), translated into possessive individualism[5] extended to identity, is the ontological premise of entering the (sacred, puritan) space of debate. The danger is that, of course, this paradise is only for the select few, for those with the "correct" identity. This also pushes everyone into identification (a question discussed in Chap. 6). The absurd but logical consequence if pushed to the limit is that the debate itself disappears, since whoever enters the debate has to come out at the other end identical to itself, and the scope of the debate becomes only that of affirming one's identity, often at the expense of others. Salvation (in puritan terms) is inscribed in the identity, identity is a sign of that very salvation, and that cannot be changed by debate. The debate then is reduced to a binary conversation about the rules that construct this sacred space: will it be based on general inclusion of fixed, unchanging identities, or on the selective exclusion of those deemed "undesirable," see "unsaved," in the moral order? The first is utopian dystopia, the end result being a space of diverse incommunicability—imagine a cage with differently colored plastic balls that bump against each other without leaving a trace. The second is current practice—societies based on exclusion of the "undesirable", deemed so by entrenched practices of identification based on superficial signs: skin color, gender, clothes, even tattoos. Both emphasize purity. Both visions slide into the "identity trap" (Mounk, 2023). In order to enter this generalized sacred space, one has to forge itself as pure identity and to have this identity recognized as worthy to enter. Impurity is dangerous, as Mary Douglas (2012 [1966]) brilliantly showed. Only the devil is ambiguous. There is no place or room for the mud when all we want is to solely enjoy the lotus.

Let us come back to digitization. The premises of its modeling, in the way it is done in Western societies, obscure the materiality of the virtual world. So the material effect of digitization is invisibilization and not dematerialization. Digitization is thus an effort to obscure the relationships of the material world, including, but not only, power relationships. It both transfers the costs of this materiality toward the

[5] See MacPherson (1964).

global South and feeds into the story of "clean, environmentally friendly capitalism." Language is capital in this process. For example, one of the major discussions in AI modeling is that of "reality drifting"—that is, when reality does not correspond (anymore) to the model on which the design of the algorithm is based, causing a malfunction. This language itself points to a possibly skewed or reversed view or "reality" as something that *should* correspond to the models we create in order to render it predictable, and, consequently, exploitable.

To finish, along with the gentrification and the gradual inscription of Bucharest among European capitals, the cables disappeared to a great degree. And so do the traces of materiality of the virtual world. Digitization promises dematerialization, and this promise is equivalent on some level with the reconstruction of the world as immaterially divine; Bucharest promises, like all other gentrified spaces, a paradise on Earth for the elected ones, with an unobstructed view to its beauty. However, as the idea of God refuses to go away, but on the contrary keeps coming back in the form of virtual space (O'Gieblyn, 2021), so do the reminders of our materiality.

Acknowledging the materiality of the virtual world has deep implications in the fields of power and knowledge:

On the one hand, it means rethinking over and over the idea of power, what is visible and what is not, and to whom. If the presence of the cables seemed normal to me, the power relations embedded in them were invisible. Power was hiding in plain sight. This indicates that the term itself, normalization, is misleading if we are to understand power as a field that creates the subject. Of course the water is normal for the fish and goes unacknowledged. Simultaneously, as long as we use the term normalization, we perpetuate the idea of a preexisting subject that could oppose the normalizing power. By only showing itself as normalizing, but not as generative, the entire field of power keeps itself selectively out of the vision, while giving the subject-in-becoming something to do, to oppose an illusory power.[6] Perhaps the solution is to acknowledge the entire field and render it visible completely—this is what would make today's power powerless: making visible the interconnections, the materiality of our being.

On the other hand, it means understanding knowledge production and intelligence not as disembodied *cogito* but as material *sum*. The question of materiality of AI sends us back toward the question of materiality of thinking—and the many forms of embodied intelligence. James Bridle's (2023) book, *Ways of Being. Animals, Plants, Machines: The Search for Planetary Intelligence*, is a wonderful exploration of this materiality and brilliantly puts it into relation with our very power of understanding and not only ours. Organisms relate to the environment in specific ways and through their materiality, developing forms of understanding that are proper and embedded into that particular relational entanglement. Ecology is

[6] This argument is similar with, and inspired by, Foucault's (1990) observations on power in "History of Sexuality, vol. 1": by showing only its repressive side and by presenting subjectivation as liberation, power obscures its mechanisms.

building intelligence, in the sense that intelligence is an emergence from the meeting of different forms and "ways of being"[7] with their environment.

Until very recently, we thought of intelligence as the capacity of doing what humans do. This is both anthropocentric and anthropomorphic. It arbitrarily places humans at the pinnacle of intelligence and privileges its definition in human terms. Thus, every other possible cognition is considered intelligent inasmuch as it takes a close to human form. However, acknowledging the materiality of intelligence opens us to accepting that there may be different forms of intelligence to which we can only have partial access (because not anthropomorphic) but their recognition would allow us to better understand our own. This, Bridle argues, applies to both organic intelligence and the artificial one, in one word the "more-than-human" intelligences. For example, it was the development of the Internet and the recognition of its material networks and their functioning that gave us the matter of thought for network theories and further for a different understanding of our own brain. At the same time, he warns both about the dangers of a narrow understanding and relating with AI, and about the impact that the material conditions of its creation, its ecology, may harbor:

> If the context of our relationships really matters so much, then it will also matter in our relationships with machine intelligence. Perhaps we should be thinking more carefully about the ecosystem in which we are raising AI, particularly the kind of aggressive, domineering and destructive forms which seem to be proliferating. That these systems are overly concerned with profit and loss, winning and losing, control and dominance, suggests that their ecological niche—the slice of environment shaping their evolution—is somewhat narrow. (Bridle, 2023: 67)

So, if we are to understand our relationship between intelligence, consciousness, and matter, where better to start than with the body?

References

Aamir, N., & Malik, A. (2017). From divinity to decoration: The journey of lotus symbol in the art of subcontinent. *Pakistan Social Sciences Review, 1*(2), 201–225.

AnthropoMass [Website]. (n.d.). Shift in balance. Retrieved December 20, 2023, from https://anthropomass.org/shiftinbalance/

Arendt, H. (2018). *The human condition [1958]*. The University of Chicago Press.

Bridle, J. (2023). *Ways of being: Animals, plants, machines: The search for a planetary intelligence* (1st ed.). Farrar, Straus and Giroux.

Coeckelbergh, M. (2020, October 6). *AI for climate: Freedom, justice, and other ethical and political challenges* [opinion paper]. Springer Link: AI and Ethics. https://link.springer.com/article/10.1007/s43681-020-00007-2

Crawford, K. (2021). *Atlas of AI: Power, politics, and the planetary costs of artificial intelligence* (1st ed.). Yale University Press.

[7] This concept resonates with Bruno Latour's actor-network theory and hybridity and with Hayles' cognitive assemblages. We will return to these later.

Douglas, M. (2012). *Purity and danger: An analysis of concepts of pollution and taboo [1966] (Routledge Classics)*. Taylor & Francis.

Foucault, M. (1990). *The history of sexuality* (Vol. 1). An Introduction. Vintage Books – Random House.

Kandeler, R., & Ullrich, W. R. (2009, July 1). Symbolism of plants: Examples from European-Mediterranean culture presented with biology and history of art [article]. *Journal of Experimental Botany, 60*(9), 2461–2464. https://doi.org/10.1093/jxb/erp166

Macpherson, C. B. (1964). *The political theory of possessive individualism: Hobbes to Locke*. University of Chicago Press.

Mounk, Y. (2023). *The identity trap. A story of ideas and power în our times*. Penguin Random House.

O'Brien, I. (2024, September 15). Data center emissions probably 662% higher than big tech claims. Can it keep up the ruse? *The Guardian*. https://www.theguardian.com/technology/2024/sep/15/data-center-gas-emissions-tech

O'Gieblyn, M. (2021). *God, human, animal, machine: Technology, metaphor, and the search for meaning* (1st ed.). Doubleday – Penguin Random House LLC.

Palinkas, L. A., & Zatzick, D. (2019). Rapid assessment procedure informed clinical ethnography (RAPICE) in pragmatic clinical trials of mental health services implementation: Methods and applied case study [study]. In *Administration and Policy in Mental Health* (pp. 1–9). U.S. National Library of Medicine (NLM): PubMed. https://pubmed.ncbi.nlm.nih.gov/30488143/

Strubell, E., Ganesh, A., & McCallum, A. (2019, June 5). Energy and policy considerations for deep learning in NLP [article]. Cornell University. https://arxiv.org/abs/1906.02243

Chapter 2
Body Matters

Those who see any difference between soul and body have neither.
Oscar Wilde (1903:6)

I taught for a few years the Anthropology of the body course to students in various cultural contexts, from international students in Istanbul's Galatasaray University to mostly Canadian students at Simon Fraser University in Vancouver. I would start and end the course by asking the students to write a couple of pages to answer the following question: "Am I a body or do I have a body?" Invariably the essays at the beginning of the course tend to predominantly argue for having a body. After 10–12 weeks of exploring a variety of theoretical and cultural perspectives on the body, from Marcel Mauss's techniques of the body to cyberpunk culture, the answers tend to shift toward "I am a body," or at least "it depends."

How do we perceive our bodies? And how does this matter in the conversation on artificial intelligence, or any form of intelligence for that matter? Ever since Marcel Mauss' seminal essay on the techniques of the body, social sciences showed unwavering interest in various aspects that link our body practices and perceptions to society, culture, and politics. The French social scientist had his interest sparked while recovering in a hospital in Alsace, a region that has both French and German population. Mauss observed that the nurses caring for him moved in different ways during routine operations or even while walking. Since the presence of a body is a matter of course in any aspect of life, many times we forget its presence and importance—its existence even. We thus forget our own existence, while reifying the dream of immateriality.

The effects of the design and use of AI on various perceptions of the human body are both omnipresent and seldom under scrutiny outside academia. This chapter's argument is twofold: It will first propose an incursion into the various instances of conceiving the human body as the driver of AI development. Second, it will look at how the human body itself and its perception are modified as a consequence of the generalization of AI use.

A. Balasescu, *Climate Change in the Age of Artificial Intelligence*, Artificial Intelligence, Simulation and Society,
https://doi.org/10.1007/978-3-031-90042-6_2

How do different conceptions of the ideal human body interact and shape AI design and application? And how does the use and abuse of AI reflect back on our understanding of the human body? It can be argued, without fear of being utterly wrong, that the design and use of AI, as that of most human technologies, refer back to the (human) body, either directly or indirectly.

Design

Let us start with the design. Both algorithms and the tools that support them, from smart phones and watches to cars or smart cities, revolve around the three major dimensions of body-technology interaction: ergonomic, experiential, and referential.

1. The physical support of the apps: the material dimension of the human-machine interaction is mostly designed using the standardization of the unmarked fully abled male body as universal referent. The size of the smart phones or watches are the most immediate examples. It is true that many objects can be adjustable, such as car seats, and most are regendered through consumer personalization—the market offers this as a postbuying option, feeding into the consumer habits. However, a colorful phone protection does not make up for its oversized screen that makes it more difficult to handle for almost half of the population. Adjustments are constantly made, of course, and new niches of operation are generated by the relatively recent realization of the male bias, not only for consumer objects in general but also in the cases of sciences and medicine.
2. The user experience (Ux) as a field of inquiry and intense research concentrates on the amelioration of the interaction between humans (bodies) and the products, ideally providing a positive adjusting feedback loop for the design of objects and apps. UX focuses on the ergonomics of experience and the feelings they generate—directly or indirectly revealing the role of the body in design. However, more often than not the user experience research is based on big data that are always already skewed. The experiential observations collected with qualitative methods are still considered problematic if not validated by data. Interestingly enough, as one of the persons I interviewed emphasized, and many working in the domain know, the type of knowledge produced by data differs from that produced with qualitative methods. The practice in many if not most of the cases is to dismiss the things we know experientially that happen because data do not confirm them: "No matter how much you insist that you see something happening, the question comes: do you have data to support this? If not, whatever you bring to the table is not taken into consideration." Or as a colleague of mine who works in system thinking recounts an exchange in a learning environment: after presenting detailed qualitative research on a subject, their colleague said: "and this is all interesting, but if you *really* want to know what is going on, you need to do a survey and collect the data."

Expecting to validate qualitative knowledge through quantitative methods is an unresolved conundrum in social sciences in general. This is not to dismiss UX research in any way. On the contrary, this domain is rich in knowledge production and certainly indispensable in the industry, bringing amelioration and continuous adaptability to the products. The only intention here is to point to two major effects: first, the standardization through data creates an *always already standardized and abstract subject (and body)* against which every user compares themself. Objects themselves are produced with this standard subject in mind. Second, the overreliance on data may in fact skew the very reality it aims to accurately represent. More on the relationship between data, facts, and reality is presented in Chap. 5.

3. The referential aspect of AI design toward (human) bodies is perhaps the easiest to grasp. The vast array of applications are, after all, human centered. They are meant, declaratively, to serve us, be it by mapping our DNA or profiling the human proteins, by tracing our movements and preferences, recognizing our speech or faces for identification, or mapping our physical daily exercises, by operating machines and all the way to creating early warnings for climate change disasters. The human body—individual or collective—is the focus of AI design, and algorithms are reflecting back to us, by design, our conceptions, and biases about human bodies. Referring back to bodies, machine learning (ML)-based systems identify, trace, classify, hierarchize, and reify (mis)conceptions about our bodies through both weighing and algorithmic reinforcement loops. Weighing is the process through which a certain amount of data is discarded because it appears to be marginal. This roughly means that events with low frequency are counted out, in other words a part of reality is erased. The classic examples are already analyzed in literature regarding bias in AI design, and some of them are emblematic: automatic selector for CVs that only choose males for engineering jobs or executive position advertisements that are more likely to be shown to male users. More so, facial recognition algorithms would fail to identify Black people as human—because the data on which they were based were undiverse and heavily skewed. Joy Buolamwini founded the Algorithmic Justice League while she was a doctoral student at MIT when she noticed that facial recognition algorithms recognized her face while wearing a white mask.[1] Her black complexion was unrecognizable or misclassified. Her thesis entitled *Gender Shades*[2] (Buolamwini, 2017) analyzes how the data sets influence the outcome and provides an argument for their diversification. She also discusses how the coding itself as an operation embeds the biases of the coder, proposing the term "the coded gaze": that preexistent biased gaze that gets embedded into the algorithm code (not only in the data).

[1] https://www.rockefellerfoundation.org/insights/grantee-impact-story/unpacking-biases-in-algorithms-that-perpetuate-inequity/

[2] https://www.media.mit.edu/publications/full-gender-shades-thesis-17/

The general belief and theory is that the more data accumulates, the less bias is generated. This is only partially true, because the algorithms themselves are biased by design, embedding assumptions about the meaning of the data. This is easily visible in search engines that associate words. In Chapter 1 of her book *Algorithms of Oppression*, Safiya Umoje Noble (2018) shows how a simple typing into Google of the words "Black girls" reveals by the ranking of the results, "a society searching" predominantly for, and being fed back misoginism, racial and sexual demeaning stereotypes. It is said that if you are not in the first three results of a google search, you do not exist. The questions are as follows: In what manner does one exist in those first three results? And what happens with all other forms of being that are hierarchized at the bottom of or altogether eliminated from search results?

The gendered, racialized, and classified faces and bodies are reified by algorithms that generate false associations and stereotypes as "objective" realities. In his book, *The Black Technical Body* (2023), Ramon Aron historically traces the entanglement of statistics as science with technological positivist conceptions of society and the history of racism. He shows how race and machine learning have a troubled relationship that dates back to the nineteenth century, with the rise of statistics at the height of the modern colonial advent. Machine learning, more often than not, reifies preexistent taxonomies, while prejudices get embedded in algorithm outputs. Benjamin (2019), Noble (2018), Wajcman (2004), and many other books deal with the question of race, body, gender, and taxonomic representation and reification in AI design, use, and abuse.

…Use (and Abuse)

Ultimately, what is the use of AI and technology? To study, help, assist, and heal but also to regulate, administer, classify, and distribute *human bodies* in the society. When it applies to the natural environment, more often than not the approach is within the agenda of managing the environment for the perpetuation of human life as defined within the parameters of the current dominant paradigm of what human means: human at the center, nature subservient. In the documentary *Our Planet*, when talking about the fact that 96% of the entire biomass on the planet is formed by humans and organisms mostly industrially produced to serve human consumption (livestock), David Attenborough states: "Nature once determined how we survive. Now we determine how nature survives." The statement is chilling and meant to be a wakeup call. It is also somehow incomplete. Formulated for the large public, it does operate within the idea of humans as subjects that act upon the world as objectified assemblage. This does undoubtedly happen at the level of our actions, and the effects of treating nature as an object are seen today. However, ultimately, nature now in survival mode will determine how we survive. Unless, of course, the colonial dreams of the techno-elite eyeballing Mars as Planet B or downloading humans into machines will become reality.

Let us return to the question of human body taxonomies and distribution. In the nineteenth century, the advancement in exact sciences—sciences that used mathematical exactitude to assert accuracy of observation about the physical world—was in full swing. Enthralled by their success and applicability, thinkers of human organization imagined that the exact sciences approach can be applied successfully to societies. The uncontested leader of the positivist era, Auguste Comte, who is considered the "father of sociology" advanced the idea of a mathematically ordered society. The regularities of human behavior, he posited, are subject to mathematical laws, and studying societies with tools from the exact sciences would reveal to us the incontestable truth about them and their governing laws. Statistics became the method of choice for the science of societies: sociology. As it happens, it is rather problematic to determine which came first, the chicken or the egg? Did societies always "behave" with mathematical precision, or did the positivist ideas about societies lead to their further organization as mathematical machines, through a thorough reorganization of space, time, and behaviors?

If not the society as a whole, at least work started to be organized on the mathematics of industrialization. None other than Babbage, the nineteenth-century mathematician who is another "father," this time of the computer,[3] advocated the organization of work on his imagined computational model based on the "difference machine" later to become "the Analytical Engine." He extended this model of calculating efficiencies from the computer to labor, following and expanding on Adam Smith's ideas of rational division of work. If value can be obtained from workers' effort, a way to optimize the extraction of value was to organize that work on the model of the Analytical Engine. He argued that:

> ...the industrial corporation could be understood as an analogue to a computational system. Just like a computer, it included multiple specialized units performing particular tasks, all coordinated to produce a given body of work, but with the labour content of the finished product rendered largely invisible by the process as a whole. (Crawford, 2021: 70)

Modernity as a process of invisibilization makes even more sense if approached from this perspective. If the industrial processes become invisible, the production of value can be transferred from the human labor to the efficiency of the process. Conversely, and through repetition and historical reiteration of this process, human workers become subjects of optimization, interchangeable pieces in the corporation-as-computer environment. Automation of workers' activity by breaking the task into repetitive movements, and optimization of the processes, becomes the universal law that generates value, and not the human efforts embedded in them, and invisible at this point. As Crawford observes, "[i]t was a strangely prophetic vision. Only in very recent years, with the adoption of artificial intelligence in the workplace, has Babbage's unusual twin goals of computation and worker automation become possible at scale" (idem.) And only that it was rather a self-fulfilling prophecy.

---

[3] The way many times Babbage's figure and legacy is presented obscures that of his collaborator, Ada Lovelace, who is in fact crucial in designing the binary-based operating systems or programs for the computers.

This vision created the premises for the rapid automation of work, and it was adopted at scale. The early attempts to create production lanes, first in the late nineteenth century slaughterhouses, and then in all industries, culminated with Henry Ford's car production lane. The latter became the blueprint of organization of production, heavily criticized and ridiculed in Charles Chaplin's "Modern Times," but adopted by virtually every corporation in the world. This blueprint not only contains the "advantage" of optimizing the work but also that of seeing human labor as a problem of efficiency that needs to be solved. The automation was not about improving workers' lives but about increasing work (and financial) output. "Babbage's idealized machinery aimed primarily to maximize financial returns [...] In a similar vein, today's proponents of workplace AI present a vision of production that prioritizes efficiency, cost-cutting, and higher profits instead of, say, assisting their employees by replacing repetitive drudge work." (ibid. 71).

The automation was (and continues to be) a process of organization of human bodies within the mathematically informed corporate giant laboratories, based on both the arrangements of bodies in place, breaking the bodies activities into minute and repetitive tasks, but also enrolling the most promisingly productive bodies and maintaining their level of potential productivity. This is not valid only for physical tasks. Thinking is a body activity, and the thinking outputs are rendered visible through physical effort. Typing reports and sending emails create physical stress.

The movements of well-being at the working place are widely spread, and more often than not benefic, given the ways in which work is organized today. But there is always the question of their intent that reveals the flipside of the story. Ideally they are about the employees' well-being. However, when combined with the methods of surveillance of the well-being, and with the insertion of nudges toward productivity, they start serving the productivity itself. When workers' "healthy habits" such as diet and running or going to the gym are monitored by *wearables* and compensated by the corporate culture, as is many times the case, and when those who "fall behind" are threatened by looming restructurings (a euphemism for firing), one may start wondering. When these habits are internalized and gain a moral undertone, and virtue is equated with being productive, the stage is set for people entering into competition with the automatic process itself. Thence, the fear of being replaced, first by more productive, healthier workers, and then by machines themselves. Perhaps it is not that machines will be so smart to take over the world but that we created a world that can easily be handed over to machines.

But how does one know that someone is healthier than others? What are the effects of the visualization of the measurements of bodies? The question "how do we perceive our bodies?" has multiple answers. But fundamentally we perceive them through the senses. AI software allows for a multiplication of perceptions of the body and affords both instantaneity and comparability of this perception. The instantaneity is given by the electronic perceptors attached to our bodies. The comparability by the translation of these perceptions into numerical expressions against a norm is as follows: ideal number of steps, ideal pulse, ideal BMI, etc. These ideals in themselves are in constant adjustment as data from multiple users accumulates in real time. It is thus possible to find ourselves normal today but slightly off tomorrow, while displaying the same numerical value for a measurement. The norm may have just shifted. So who, or what, are we to believe? Our senses that may tell us we feel good, or the digitized sensorial apparatus that is sold to us as an exponent of technological advancement and objectivity of body representation? We mathematically *ordinate* the bodily functions, and this process creates *the subordinated body*. Our bodies become a traceable aggregate of data regarding our caloric intake, daily steps, minutes of electronically enhanced mindfulness, muscular mass, or heart rate while on a treadmill. This perception, shaped by the standards embedded in apps from wellness to the health industry, reorders and reforms the relationship with our bodies.[4] On the one hand, our bodies become nodes of generating data for the electronic economy that brings the healthcare and well-being industries under its umbrella. On the other hand, we start perceiving our bodies as vehicles of constant improvement; resources to be exploited in our attempt to "grow," get a better job, a better partner, or self-reflexively an improved state of well-being. The physiology of the body morphs into the economy of constant self-improvement that reflects back on productivity.

This economy comes to a sudden stop when facing death. AI and biotech say it does not need to be so. The perception of finitude as a problem is a cultural particularity of Western modernity. During the short history of modern thought and scientific practices, this perception has generated endless efforts to overcome death. The latest attempt at immortality is held by the promises of AI to transcend the human body and download consciousness into a machine, as in the cyberpunk novels by Greg Egan, the blueprint of today's singularity embraced by the likes of Kurzweil, while biotech promises the possibility of eternal rejuvenation of that same body.

The most prominent exponent of the current credo in conquering death is Bryan Johnson, the Silicon Valley billionaire who put his money behind himself as a project of immortal human, creating the movement "Don't die." He is using a patchwork combination of AI algorithms, bioengineering, and medicine including transfusions of bodily fluids from his son, dieting and other techniques in order to perpetually rejuvenate his body, or at least this is his claim. Listening to his interview in Trevor Noah's podcast "What Now," one can grasp both his philosophy and his methods. I will briefly dwell on the latter: his approach is "to measure everything that can be measured in the body" parameters, including his nighttime erections in terms of intensity and duration and aiming to ameliorate these measurements constantly. I found the erectile measurement particularly interesting both because it points out to masculine standardization of measurements and because it is also a reminiscent of the aspirations of Immortan Joe, the fictional character in the postapocalyptic film *Mad Max 4, Fury Road*. His idea of achieving immortality was by impregnating the most eligible women in the land he controlled. The metaphor could not be clearer.

---

[4] It is clearly difficult to escape the language that asserts the existence of an external-to-the-body subject. It is perhaps not even important to do so. What is necessary is to integrate in our thinking the dynamicity of the self and the acceptance of its continuous subjectivation. Thus, the relationship body-self is not a unilateral one but a circular, reciprocal one. The whole forms an aggregate.

The movie is a complex critique of a certain form of patriarchy, colonialism, and extractive capitalism.

Of course the general measurement, weighing, and scheduling of the body also facilitate the use of AI, because the tool itself is based on numerical data—it will become clear later in the book why this is crucial. The life schedule is also strictly measured and followed, aiming to the unique goal of achieving immortality, for him, for his son, and also for his father. It is interesting to remark that in his photo advertising, he appears sometimes with his father and son together, and in the podcast, he never mentions his mother. Of course this also has a commercial side to it. The Bryan Johnson Blueprint being a brand that sells one the techniques developed by him to achieve, well, at least a prolonged, if not endless, life. It is perhaps worth mentioning that Bryan Johnson is a member of the Church of Latter Day Saints and he spent 2 years as a missionary in Ecuador between 19 and 21 years old. His figure made headlines in the past and current year, and publications such as Vanity Fair and The Atlantic ran feature length articles about him. When I first saw his photo, I was impressed. I thought he was a person somewhere between 55 and 60 years old, managing to look in his 40s. I was somehow disappointed to read that he is only 47.

The seduction of immortality by achieving optimal, measurable parameters ultimately creeps into our handheld devices through the apps that we download compulsively, use obsessively, and seldom discard even if we happen to forget about them. This is a new form of subjectivation, more insidious, but at the same time, it renders the mechanisms of power and subjectivation easier to understand. If the use of electronic record systems in healthcare generates new relationships with the technology and thus a new "kind" of healthcare worker, so does the use of wearables that "sense" our bodies and retransmit the information to our eyes under numerical form. Senses are not only a means to sense oneself, but also a means to think (of) oneself, and gain self-consciousness, as Merleaux-Ponty (2013) emphasized in his writing on the phenomenology of perception. The senses are the constitutive part of both the thinking about oneself and in thinking the self. The visualization of bodily function, thus, generates new subjectivities by turning inside out the means of subjectivation, while simultaneously internalizing them.

The disintegration of physical bodies in measurable parameters and their reintegration in the data cloud reshapes our understanding of the body/self-aggregate, and the perception of the physical world at large, speaking volumes about our relationship with the environment. In so many ways, the body as we understand it is a political project. And reclaiming the body as ourselves is also a political project, most prominent in feminist movements starting in the second part of the twentieth century.

Biopower

In the last chapter of *The History of Sexuality*, Michel Foucault (1990) theorizes "biopower" as the form of power that governs modern society. He describes it in contrast with "sovereign power" or the power over death that governed during the "classic period."[5] The main expression of sovereign power is the capital sentence, as death is the only way in which the sovereign could express and display his power over his subjects. Foucault argues that along with a series of transformations in everyday practices, and a new conceptualization of the social body, biopower emerges as a new system of regulation for life—both individual and collective:

> Power over life evolved in two basic forms; these forms were not antithetical, however; they constituted rather two poles of development linked together by a whole intermediary cluster of relations. One of the poles—the first to be formed it seems—centered upon the body as a machine [...] ensured by the procedures of power that characterized the disciplines: an anatomo-politics of the human body. The second, formed somewhat later, focused on the species of body, the body imbued with the mechanics of life and serving as the basis of social processes: propagation, births and mortality, the level of health, life expectancy and longevity, with all the conditions that can cause these to vary. Their supervision was effected through an entire series of interventions and regulatory controls: a bio-politics of the population. (Foucault, 1990:139)

The appearance of the anatomopolitics of the human body forms the domain of the microphysics of power, and it is extensively treated in "Discipline and Punish" (Foucault, 1979). The constant exercise of power over the body creates its own subject: the individual. Foucault defines discipline as "methods which made possible the meticulous control of the operations of the body, which assured a constant subjection of its forces and imposed upon them a relation of docility-utility" (Foucault, 1979:137). From the end of the seventeenth and throughout the eighteenth century, while the social body was conceptualized by philosophers as the aggregation of independent, monadic subjects-participants in the "social contract," discipline was a practice that pervaded various domains of life, in schools as well as in the family, in the army as well as at the workplace. Discipline has a major role in the "practical" creation of individuality, while the theoretical role belongs to philosophers:

> The individual is no doubt the fictitious atom of an 'ideological' representation of society; but he is also a reality fabricated by this specific technology of power that I have called 'discipline' (ibid. p.194).

The striking difference between the "ideological individual" and the "individual as product of discipline" is the status of the body. For the former, the body is a possession upon which she/he can exercise the will, while for the latter, the body is always already possessed (because of being created) by and through the exercise of micropower. The exercise of this power forms its own "objectified subject." In order to operate with the form of power described, it is also necessary to rethink and

---

[5] Foucault identifies the classic period as the historical moment that finishes around the French Revolution.

moreover renounce the idea of separation between subject and object. The focus of any analysis would shift thus from the subject to the practices of subjectivation.

Foucault is tracing the genealogy of the individual, the human being as we relate to it in the modern order of things, from the relations between body and surrounding instruments, tools, and spaces both constraining and enabling for the movements, be they physical or mental. He intentionally chooses the example of imprisonment and torture, for reasons both of political engagement and literary effect. Foucault's examples are strikingly visual, one can easily imagine the torsions of the bodies under the instruments of the executioner, or the confinement of young pupils into their school benches. The double relationship body-object/body-subject is of central significance. Inasmuch as objects directly influence movements, attitudes, and body positions, and subjectivity is the result of bodily motions, one has to reexamine the subject in its relation with surrounding objects. Or, as Jean-Pierre Warnier better formulates the matter:

> [...] the word 'bodily' induces a distinction between the body and something else, that would be the subject. Or a subject does not "possess" a body. It is a body. Talking about the motion conduits [conduites motrices] of a subject we avoid the trap of the dualism hidden behind the vocabulary of the body. (Warnier, 1999: 10, my translation and italics)

Warnier provides a basis for a theory of the subject, through a rethinking of the very relation between subject and objects. Warnier places the objects (material culture) at the center of this theory and in their relation with the body, that is, with the subject. The mediators between objects and the subject are the "motion conduits," a term that the author prefers to Mauss's "body synthesis" (Mauss, 1992 [1934]) for reasons explained above.

> 'Body synthesis' or 'motion conduits' refer to the integration of movements necessary to operate or use any object; this is not always a conscious process, but it is omnipresent in the dynamic rapport that the body has with the surrounding material world, and with itself. In both cases [motion conduits vs. body synthesis, my note] we deal with the same phenomenon, that is to know the capacity of the subject to memorise or integrate motion conduits perfectly adapted to the dynamic of the relation with the objects and the environment—conduits, meaning the assembly of finalized motion actions. (idem)

In fact it is this very dynamic that interests Warnier. There are two main reasons for this: firstly, the author is able to review anthropological theories regarding human evolution, and secondly, because of the relevance that access to, and the use of, objects has for the socially positioned subject formation. In other words, power and social order emerge from the body-material culture interaction, while "the subject" is in a permanent production and does not preexist as an individualized actant.

Certainly this collapse of distinction subject/object has multiple implications, many discussed at length in a diversity of academic fields, from anthropology to science and epistemology. Most importantly, it undermines the Cartesian edifice of modernity and the sociocultural order it imposes, by questioning its fundamental assumption: that the individual exists as a disembodied thinking entity who can take possession of the surrounding world, starting with *its body*. The way we relate to AI is based on this vision, and AI is viewed by the proponents of technooptimism as a possibly better, because virtually disembodied, thinking entity: the perfect subject, perfectly objective because everything outside it is by definition an object of its encoding gaze. There is a paradox inside the field of AI itself, and there is a dispute there, too. A part of the representatives of the field are strongly Cartesian, imagining that AI is only a matter of coding and digitization, and general intelligence may be achieved in the so-called virtual space. Neural networks and their advancement brought back the hard problem of disembodied intelligence: the body and AI builders know it. Neural networks, because they imitate the brain, are thought to be the possible solution to arrive at this ideal form of intelligence. But once this conversation is opened in the field, the body creeps back in. Here is Yann LeCun, the head of AI at Meta, talking about it:

> On the whole, current A.I. technology is talky and cerebral: it stumbles at the borders of the physical. 'Any teenager can learn to drive a car in twenty hours of practice, with hardly any supervision,' LeCun told me. 'Any cat can jump on a series of pieces of furniture and get to the top of some shelf. We don't have any A.I. systems coming anywhere close to doing these things today, except self-driving cars'—and they are over-engineered, requiring 'mapping the whole city, hundreds of engineers, hundreds of thousands of hours of training.' Solving the wriggly problems of physical intuition 'will be the big challenge of the next decade,' LeCun said. Still, the basic idea is simple: if neurons can do it, then so can neural nets. (Rothman, 2023)

What is missing in this quote (and in this type of thinking) though is the fact that biological neural nets are part of the larger, open system, the body, who is in turn part of the larger open ecosystem and so on. To be really aspiring to general intelligence, artificial neural nets would need to jump out of the enclosed limitations of the network and out of the closed thinking projected upon them, namely, *that they can be a more perfect knowing and bounded subject in, and of, an objectified external world.* That this world exists independent of the knowing subject is clear, but that the knowing is independent of the subject's relation to the world, much less. In the late 1940s, von Uexküll (2010 [1934]), a German biologist, published an essay entitled *A Foray into the Worlds of Animals and Humans. With a Theory of Meaning* in which he explores the environment as the production of meaning—generated by, and useful to, the subject for survival purposes. He postulated the continuity of the environment and the subject that form a circular relation: the subject (the organism) perceives the meaningful environment for its survival and generates the meaning of that perception by the virtue of being constructed by and through this meaning-assigning system:

> All the organs of plants as well as of animals owe their form and their distribution of materials to their meaning as utilizers of the meaning factors which come to them from the outside.
>
> *The question as to meaning must therefore have priority in all living beings.* Only once it is solved does it make any sense to research casually conditioned processes, since the activity of living cells is directed by their self-tones (Uexküll, 2010 [1934]:151, original italics)

To synthesize, von Uexküll quotes Goethe "Were the eye not sunlike,/ It could never gaze upon the sun" (ibid. 190, original italics). His theory of meaning and its centrality in the making of the knowing/living subject is reprised by poststructuralist thinkers precisely in order to deconstruct the assumption of the distinction knowing subject/knowable objectified world.

For the purpose of the proposed analysis, I will retain three of the more specific implications of the rethinking of subject/object distinction in terms of subjectivation that eliminates the precedence of the (thinking, individual) subject.

First, there is the potential of a better understanding of the critique of technologies as political means of subjectivation, repressive and productive simultaneously. The way technology is produced and represented is a political act—because an act of subjectivation, of the creation of the subjects that enable the exercise of power itself. This includes popular culture—films, books, music, etc.—but also instruction manuals or the way in which one writes about technology in news outlets and magazines. Technology thus is not neutral nor objective in its applications or in its outcomes. From the gross characterization of tech as enabler of life or tech as evil force of subjugation, to the subtler representations of technological, gendered, and racialized bodies, pop culture shapes perceptions about human bodies, their abilities, and their sociocultural role. It also informs, more often than not, the way in which humanoid robots are designed in real life and used.

The masculine warrior body, the feminine subservient, emotional, and oriented toward pleasing men, and the predominantly white faces associated with AI-endowed robots in publicity, they all subtly contribute to reinforcing stereotypes. For example, with few exceptions, male-lover robots are quasiabsent in pop culture. One needs to dig into high-brow literature, such as the book *Machines Like Me* by Ian McEwan (2019) to find them. The rest, starting with "Blade Runner" all the way to "Her" and "Ex-machina," pleasure-inducing machines are female, suggesting that pleasure itself is the domain of masculine dominion and entitlement. This fits very well with the patriarchal view of sexuality. Kathleen Robinson writes extensively about the real impact on gender expectations that the perpetuation of this trope has (Richardson, 2017). Also, in the current patriarchal context, it would be hard to imagine a proliferation of images of AI or robots with male characteristics that make human women fall in love or obtain pleasure from them.[6] It is bad enough that "they" are taking "our jobs," now are they after "our women," too?! The fear of redundancy and obsolescence of "traditional" masculinity is felt all the way into cyberpunk. Or maybe it is where it starts.

Second, there is the anthropological and phenomenological understanding of the matter of consciousness or what LeCun calls "physical intuition." The subjectivation theory proposed by Warnier implies that the body is an integrative part of the thinking process. The very subtitle of his book is *The Human Who Thinks with His Fingers*. Through advancing the concept of motion conduits, Warnier explains both the continuous subjectivation process in relation to the material culture and the

[6] Again, there are exceptions as the McEwan's book mentioned here. One of the themes of the book is precisely the "amorous disorder" that a male android may provoke.

process of thinking as an emerging phenomenon from the interaction between human-body-as-subject-in-becoming and the surrounding world.

But what about other bodies? Will they generate "other minds"? Peter Godfrey-Smith (2016) is certain about this, and he convincingly argues that a radically different biomorphology, like that of the cephalopods, shapes radically different forms of consciousness and minds. The Australian philosopher closely studied octopuses and built the argument that the consciousness and thinking of radically different bodies develop divergently and reflect that difference to the point of being unrecognizable. If we more readily accept that primates are intelligent, and to a certain extent any mammals, we do so because they have relatively similar bodies with ours that potentially generate similar minds. This recognition happens at a subconscious level. Conversely, body-mind assemblages like those of cephalopods display too big of a difference from us in order to recognize them as intelligent at a first sight. Moreover, there is also the difference in the environment of acting and developing, that is, the becoming of the mind. The ocean's depths offer a completely different interface between the thinking body and its surroundings, shaping different perceptions from humans not only due to the anatomophysiological differences but also due to the quality of the space of immersion and time. Whoever does scuba diving knows that once under the sea, the entire universe of feelings changes, including the perception of time. It is not only a matter of speculation to affirm that time is not only differently perceived but also gains a different meaning *in* organisms living perpetually *under* the sea—accepting a different understanding of "meaning" than that of "something" assigned by the preexistent meaning making subject. This immediately begs the question: is it the body alone or bodies together? Are there only human bodies, or every kind of body, including technological ones?

Tackling these questions we can move further away from *the thinking body toward the body in (the) thinking,* and this is perhaps a different way to define the idea of subjectivation: the continuous process out of which thinking emerges, comprising the system of objects and environment that surround the subject in becoming. Out of this conception, the question emerges: who does the acting? What then is "agency" that capacity we traditionally assign to the individual subject? The actor network theory (ANT) advances the idea that agency is distributed among what we traditionally think of as subjects and the material world. The latter does the thinking by determining the material conditions out of which the thinking emerges. The self-conscious subject is constantly created and actualized in the moment in which thinking is expressed (through words or actions); however, it is this very emergence of thinking out of the system that creates the subject which does not necessarily exist in that form prior to the process. One of the well-known proponents of this theory is Bruno Latour, whose work body of influenced various domains from political anthropology to economic thinking. Applying it, a different understanding of the meaning of economical market emerges, as presented by Callon and Muniesa (2003), whose work implications were analyzed by me elsewhere together with a finance specialist:

> The authors postulate that we can only understand the market as a collective computational device that crunches all three elements if we do away with the split between human agents and their products. They propose an understanding based on co-production that helps us move beyond the separation of subject/object and facilitates a deeper understanding of both objectification and singularization—in other words the calculability of sale-able things, or goods (that are not necessarily tangible objects). In their view, the market is a distributive calculative agency, formed by multiple actants that assign prices to goods. (Balasescu & Jain, 2018: 8)

This example illustrates the decentering of the idea of agency from the human individual or collectives and also makes clearer what is a "cognitive assemblage": a system comprising the material and nonmaterial elements and their interlinks out of which cognition emerges (Hayles, 2014). Practically, not only can a cognitive assemblage be anything, but also, theoretically, anything can be a cognitive assemblage, given that agency has a means of expressing itself.

The cognitive assemblages may scale up to any dimension. Alexander von Humboldt was a pioneer, and a contrarian of his era in Western modern thought, when he came to understanding living systems as thinking systems. He is the precursor of Gaia theory, and, right in the historical moment in which Cartesian philosophy was taking hold, he argued that thinking is not something necessarily and solely associated with the individual's brain, be it human or otherwise. He envisioned the entire planet as a thinking organism in his book, *Cosmos* (more on his work in Chap. 4). This idea that emerged in many forms offers an extended vision of bodies beyond the individual. It may or may not have at its center human consciousness as a form of thought expression.

Planet.com is a company that sends out into orbit mini satellites equipped with cameras that photoshoot Earth. The company provides "global monitoring and analytics" about our planet, basically a visual continuous charting of the Earth that may identify continuities and change on its surface. Clients of this company comprise businesses and civil governments from both the USA and Europe. A couple of years ago I interviewed one of the founders, Andrew Zolli, for this book. The implicit intention of the company is to increase the "self-awareness of the Earth," he told me. The cameras are, in his words, "the way to create a consciousness for the planet." In other words, in his perspective, this is how the planet senses itself, à la Merleau-Ponty. This sensing however is mediated through human vision, but Andrew sees the cameras, and us for that matter, as akin to sensing organs of the planet, too.

A comprehensive recent book on these topics, indeed a practical meditation on many possibilities of consciousness and minds, is the already mentioned *Ways of Being. Animals, Plants, Machines: The Search for Planetary Intelligence* by James Bridle (2022). The book is filled with stories and examples of multiple types of cognitive assemblages, from map-creating mold in Japan to mechanical turtles or computing water streams. One of the most important arguments of the book is that concerning the computational metaphor that permeates the current thinking about the world: that the world functions as a computer. This metaphor is of course afforded by the prevalence of technology today and by the recent proliferation of language surrounding them. Bridle turns the argument on its head: computers are like anything else in the world. They are part of it, part of cognitive assemblages, and not supercognitive entities (real or imagined). They can scale up the process of thinking, and they are part of it, together with humans and the environment we created for their learning. But they are not monadic subjects that can understand an objectual, objectified, and thus objective world. Neither are we. And because they are like the world, they display its limitations. Biology may be one of them. Let us briefly explore how.

As easily as we can scale up cognitive assemblages, we can also scale them down. Scaling down simply means to disassemble something that we think of as unity, in order to understand it as assemblage itself. Michael Levin, a researcher at Wyss Institute for Biologically Inspired Engineering at Harvard University, proposes a framework of understanding of the phenomenon of thinking based on the principles of scaling up, scaling down, and unification of thinking into what he calls a "holobiont": an assemblage of multiple agents that is truly more than the sum of its parts. The organism results from a form of anatomical homeostasis of these scaling processes: for example us humans, a holobiont made out of cells, bacterial colonies, viruses, known as the microbiome. "[...a]natomical homeostasis can be viewed as the result of the behavior of the swarm intelligence of cells" (Levin, 2022: 1). The elements whose systemic assemblage forms the mind in questions are called elsewhere by Levin "agental material" (Levin, 2023).

There are multiple implications of this framework of understanding on the theory of minds and selves that the author calls TAME (Technological Approach to Mind Everywhere), in evolutionary theories, in philosophy of minds, and in the politics of approaching different forms of intelligence, including the artificial ones. One of them is the proposition of an indistinguishable border between artificial and biological intelligence. The author argues that with this border, it becomes irrelevant if we think in the terms of TAME:

> For example, one view is that only biological, evolved forms have intrinsic motivation, while software AI agents are only faking it via functional performance [but don't actually care (Oudeyer and Kaplan, 2007, 2013; Lyon and Kuchling, 2021)]. But which biological systems really care—fish? Single cells? Do mitochondria (which used to be independent organisms) have true preferences about their own or their host cells' physiological states? The lack of consensus on this question in classical (natural) biological systems, and the absence of convincing criteria that can be used to sort all possible agents to one or the other side of a sharp line, highlight the futility of truly binary categories. Moreover, we can now readily construct hybrid systems that consist of any percentage of robotics tightly coupled to on-board living cells and tissues, which function together as one integrated being. (Levin, 2022)

This goes to the heart of the dispute around human augmentation that has at its center the question of "how much" a human organism can be augmented and still remain human. Some say not too much, others that it does not really matter, since the mind is distributed and there is no such thing as a subject, only subjectivation. This of course will beg the question of authenticity and its quality of signaling virtue in society (Iftode, 2019; Iftode et al., 2022). We can extend the question here to "how much organic matter do we need to accept an assemblage as intelligent"?

I am of two minds around these questions: on the one hand, I agree that moving away from the subject/object division is a necessary step toward understanding other minds and accepting other forms of intelligence. On the other hand, I would not hand over the "intelligence" as capacity to be conscious to purely inorganic machines. While I agree with Levin concerning the necessity of doing away with a general sharp line between what is intelligence, and what is not, I would argue that there is a sharp line not in the boolean scale of quantity of elements until an assemblage becomes intelligent (Buckner, 2024) but in the scaling down of the elements that constitute it. Keeping with his technological metaphor (itself problematic) if we dismantle a biological organism, each of its components is also a mechanism, and we can continue to dismantle it, finding mechanisms within mechanisms, within mechanisms, all the way to subcellular levels and DNA. If we dismantle a human-made computer, we will very soon arrive at a hard stop, the mechanisms within mechanisms disappear, leaving the place to mineral or artificial materials: plastic, silicon, glass, etc. So the limit is not in scaling up and combining biological with nonbiological agental material to form new forms of intelligence but in scaling down, when we reach a monad "subject" that has a limited or no cognitive "mechanisms" embedded in it such as an individual computer. And this is perhaps the sharp line of caring (as in "does a fish care as opposed to a machine?"): not the intelligence as real or as an imitation act but the capacity of the assemblage—or of a part of it at least—to suffer or to love for that matter.

Here we enter a different domain, and the dispute can be infinite. However, I think we are not really wrong if we say that an assemblage of solely mechanical elements, with no organic agental material, is not suffering, despite the capacity of displaying signs that we may interpret as such. It can be programmed to do so, and we are suckers for empathy. The weird and worrisome characteristic of today's world is that we are more prone to accept machine suffering than that of a fellow being, human, or otherwise.

Suffering aside, the careful moving away from binaries, the gradual approach to intelligence and other minds, as Levin himself advocates, is both necessary and not without troubles. One of its important aspects is that it problematizes the idea of individual ownership, authorship, and copyright; indeed the very fabric on which the modern mythology of the human creator is based and with it a big part of today's economy. The shifting away from this idea impacts both individual creators (and the screenwriters strike in 2023 is the most recent symptom of this impact) and the way we organize our economy of creation but also our regime of truth as produced and communicated through writing. For a comprehensive conversation on these aspects, Foer's book *A World Without Mind* (2017) is a good starting point. Briefly, one of his concerns—that proved to be absolutely justified—is that by doing away with the myth of the creative subject, we can easily justify an economy that refuses to pay the human subjects for their creative work, that is, the authors of cultural products. We have seen it happen. The system starts paying the public "engagement" through social platforms and not the effort of creation itself. When myths are dismantled (and Foer is very clearly aware that we are talking about myths), we need to be ready with something else to propose in its place or at least be aware of the damage this dismantling does. "Moving fast and breaking things" is only a half-thought-out attitude.

The third implication of the collapse of subject/object distinction, indeed central in the way we think about life, has to do with the large body of theories and practices that fall under the umbrella of feminism. Moving away from subject/object separation indicates toward feminist theories of the body and subjectivation *not as theories of, and about, the subordinate "other" (other bodies/other subjects than the subject by excellence, the unmarked male body) but as central for the understanding of any living body,* that is, of anybody alive or, better said, of any alive body.

A seminal work on feminism and cyberbodies is, of course, Donna Harraway's *Simians, Cyborgs and Women. The Reinvention of Nature* (1991), a series of essays written mostly in the 1980s, in which the author explores the multiple possibilities of bodies and minds, once we gave up on the standard subject. She sustains the feminist idea that nature as we find it in the moral concept of objectified reality for the knowing subject is the privileged construct of patriarchy. The postulation of the limit subject/object, rational/emotional, humanness/wilderness, allows the ideal subject (the male) to take dominion of the objectified world, women included. This is a theoretical direction well known in deconstructionist approaches on binaries that define (Western) patriarchal culture. The fascinating path Harraway takes is that of the avoidance of the solution of "return to nature" that will only reinforce the binary. This places her among the pioneers in thinking differently than the norm about technology, since she places technology within the realms of possibilities of breaking the binary thinking and not of reinforcing it. The cyberorganism, the biotechnological composite, "the freak" outside the norm, as she calls it, is instrumental in dismantling the binary. She advocates thus the necessity of integrating feminist theories within sciences, and not away from them, pointing out how the rejection of science in developing gender theories only reifies the binary:

> We have perversely worshiped science as a reified fetish in two complementary ways: (I) by completely rejecting scientific and technical discipline and developing feminist social theory totally apart from the natural sciences, and (2) by agreeing that 'nature' is our enemy and that we must control our 'natural' bodies (by techniques given to us by biomedical science) at all costs to enter the hallowed kingdom of the cultural body politic as defined by liberal (and radical) theorists of political economy, instead of by ourselves. (Harraway, 1991:9)

This quote encapsulates both the issue with scientism, that is, the idea that there is a separation between "exact" sciences and the rest, linked to the question of method, and the centrality of the body in any forms of political and cognitive organization.

The binary separation has it that on the one hand science is double: hard science (so-called exact)/soft sciences and science/culture in the larger sense. This allowed the assertion that only positivist quantitative methods legitimize *real* scientific approaches (a question I will deal with in Chap. 4) and on the other hand to postulate the hierarchies of societies based on scientific versus "magic/primitive" worldviews. Since the beginning of modern anthropology, the latter is questioned by Malinowski in *Magic, Science and Religion* (2015 [1948]). He refutes forcefully his contemporaries' view on "primitive" societies as living exclusively in an "inferior" realm of superstition and enchantment, although he does it in his era's language (how else?):

> There are no peoples however primitive without religion and magic. Nor are there, it must be added at once, any savage races lacking either in the scientific attitude or in science, though this lack has been frequently attributed to them. In every primitive community, studied by trustworthy and competent observers, there have been found two clearly distinguishable domains, the Sacred and the Profane; in other words, the domain of Magic and Religion and that of Science. (Malinowski 2015 [1948]: 17)

In doing so, Malinowski shows how hierarchies of dominance are constructed based on assertions of nature as object of science and on the hierarchization of those societies that allegedly lack science and thus are presumably closer to "nature," rendering them "savage" and placing them under the "legitime dominion" of the science-based societies. These days this type of dichotomy appears in the language of separation and assertion of "cultural differences," in which difference is thought of in relation to an unmarked, implicitly superior self. This language is pervasive; we find them from pop cultural products all the way to supermarket aisles: from "world music" (implying that there is some music outside and above the world) to ethnic food (implying that there is a nonethnic standard) we encounter a whole array of expressions that indicate to an aspirational, unmarked subjectivity—the nonethnic (absurd as it may sound), unmarked, "universal" subject. The return into our bodies, and the acknowledgment of the multiplicity of aggregate, scaled up, down and lateral bodies, and implicitly that of the multitude of possible minds, would allow us to move away from dominance into a domain of freedom that may be threatening. I will make the argument that the threat felt is death itself, and the clinging to dominance is the desire to dominate one's own finitude. Meanwhile, let us shift our focus to the matter of science and culture guided by another paragraph from Donna Harraway's work:

> In evading the importance of dominance as a part of the theory and practice of contemporary sciences, we bypass the crucial and difficult examination of the content as well as the social function of science. We leave this central, legitimating body of skill and knowledge to undermine our efforts, to render them utopian in the worst sense. Nor must we lightly accept the damaging distinction between pure and applied science, between use and abuse of science, and even between nature and culture. All are versions of the philosophy of science that exploits the rupture between subject and object to justify the double ideology of firm scientific objectivity and mere personal subjectivity. This anti-liberation core of knowledge and practice in our sciences is an important buttress of social control. (Harraway, ibid.: 8)

References

Bălăşescu, A., & Jain, A. (2018). Financial bubbles and their magic. *The Journal of Philosophical Economics, XII*(1), 1–35.

Benjamin, R. (2019). *Race after technology: Abolitionist futures*. Polity Press.

Bridle, J. (2022). *Ways of being: Animals, plants, machines: The search for a planetary intelligence* (1st ed.). Farrar, Straus and Giroux.

Buckner. (2024). *From deep learning to rational machines. What the history of philosophy can teach us about the future of artificial intelligence*. Oxford University Press.

Buolamwini, J. (2017). *Gender shades: Intersectional phenotypic and demographic evaluation of face datasets and gender classifiers* (MIT Master's Thesis). Mit Media Lab.
Crawford, K. (2021). *Atlas of AI: Power, politics, and the planetary costs of artificial intelligence* (1st ed.). Yale University Press.
Callon, M., & Muniesa, F. (2003). Les marchés économiques comme dispositifs collectifs de calcul [Economic markets as collective calculation devices]. *Reseaux, 21*(122), 189–233.
Foer, F. (2017). *World without mind: The existential threat of big tech* (1st ed.). Penguin Press.
Foucault, M. (1979). *Discipline and punish: The birth of the prison* (1st ed.). Vintage Books – Penguin Random House.
Foucault, M. (1990). *The history of sexuality, Vol. 1: An introduction* (1st ed.). Vintage Books – Random House.
Godfrey-Smith, P. (2016). *Other minds. The octopus, the sea, and the deep origins of consciousness* (1st ed.). Farrar, Straus, and Giroux.
Hamilton, M. (2020). *The coded gaze: Unpacking biases in algorithms that perpetuate inequity*. The Rockefeller Foundation. https://www.rockefellerfoundation.org/news/rockefeller-foundation-announces-6-million-additional-funding-100000-inner-city-students-nationwide-see-hamilton/
Harraway, D. (1991). *Simians, cyborgs and women. The reinvention of nature* (1st ed.). Routledge.
Hayles, N. K. (2014). Cognition everywhere. The rise of the cognitive nonconscious and the costs of consciousness. *New Literary History, 45*, 199–220.
Iftode, C. (2019). Assessing enhancement technologies: Authenticity as a social virtue and experiment. *The New Bioethics, 25*(1), 24–38.
Iftode, C., Zorilă, A., Vică, C., & Mihailov, E. (2022). Experimental and relational authenticity: How neurotechnologies impact narrative identities. *Phenomenology and the Cognitive Sciences*, 1–18.
Levin, M. (2022). Technological approach to mind everywhere: An experimentally-grounded framework for understanding diverse bodies and minds. *Frontiers in Systems Neuroscience, 16*, 768201. https://doi.org/10.3389/fnsys.2022.768201
Levin, M. (2023) Darwin's agential materials: evolutionary implications of multiscale competency in developmental biology. *Cellular and Molecular Life Sciences, 80*, 142. https://doi.org/10.1007/s00018-023-04790-z
Malinowski, B. (2015 [1948]). *Magic, Science and Religion*. Martino Publishing, Mansfield Centre, CT.
Mauss, M. (1992 [1934]). Techniques of the body. Incorporations. In J. Crary., & S. K. Winter (Eds.), *Zone Books* (pp. 455–479).
McEwan, I. (2019). *Machines like me: A novel* (1st ed.). Penguin Random House LLC.
Merleau-Ponty, M. (2013). *Phenomenology of perception*. Routledge – Taylor & Francis Group.
Noble, S. U. (2018). *Algorithms of oppression: How search engines reinforce racism*. NYU Press.
Ramon, A. (2023). *The black technical body: On machine learning and the aspiration of black being*. Sternberg Press/The Antipolitical.
Richardson, K. (2017). *An anthropology of robots and AI. Annihilation anxiety and machines*. Routledge.
Rothman, J. (2023, November 20). *Why the godfather of A.I. fears what he's built*. The New Yorker. https://www.newyorker.com/magazine/2023/11/20/geoffrey-hinton-profile-ai
von Uexküll, J. (2010 [1934]). *A foray into the worlds of animals and humans. With a theory of meaning*. University of Minnesota Press.
Wajcman, J. (2004). *TechnoFeminism*. Polity Press.
Warnier, J.-P. (1999). *Construire la Culture Materielle. L'Homme qui Pensait avec ses Doigts [Constructing material culture: The man who thought with his fingers] (French)*. Presses Universitaires de France.
Wilde, O. (1903). *Phrases and philosophies for the use of the young*. [Digital archive] Retrieved November 1, from https://web.english.upenn.edu/~cavitch/pdf-library/Wilde_Phrases_and_Philosophies.pdf

Chapter 3
Culture Matters

Automation systems are not designed, and do not operate, in a void. They are created in places with a specific culture, and they are spread around the world more often than not with the expectation that they will work as intended. Usually though, they tend to work as designed. In contexts other than those for which they were designed (and I refer here not only to the labs), the automated systems based on models may easily malfunction (as we saw in the first chapter) or entirely fail to give the expected results. More so, as Bridle (2022) observed, the environment and tasks that get embedded in the training settings; the overall emphasis on competition, gain, and exploitation; and the stories that precede the algorithms' making and that get amplified in their use are determinants for the shape of this tool and of the world we imagine while using it.

Again, it is not the system per se that fails but our imagination; the failure is the assumption that the interaction of the acting model with a given reality will give the same results, irrespective of how the reality presents itself. They are not only to be adopted but also to be adapted. The success of AI adoption depends on the web of relationships between the agents that produce, develop, redistribute, and share knowledge: humans and machines alike, as they exist in the overall environment. When designing an object, AI systems included; the culture surrounding their creation gets partially embedded in them. It gets transmitted through these objects but also changes along with their use.

What cultural biases get encoded in the algorithmic models, and can they be culturally redesigned? Why not consider how the Silicon Valley culture of programmers gets transmitted through algorithms and how they influence every other place that they touch? Why not think about the imposition of forms of thinking through the logic of algorithms? Can we discuss algorithmic coloniality? These are far from being new questions. However, this chapter proposes a perspective that takes into consideration the diversity of both individuals and cultures and proposes a critical approach to the possible blindspots that the very definition of "culture" creates

A. Balasescu, *Climate Change in the Age of Artificial Intelligence*, Artificial Intelligence, Simulation and Society,
https://doi.org/10.1007/978-3-031-90042-6_3

when attempting to design "culturally sensitive" algorithms. I will start by discussing again examples generated at the meeting of algorithms with human bodies in the field of healthcare, in order to move the conversation toward the larger aspects of culture. Let us start with setting the stage.

Optimization is a major selling point of artificial intelligence (AI)-based automation systems. It consists of clear and obvious applications—such as optimizing the image-based diagnosis in healthcare. All the same, it bears promises of a perfectly optimized lifestyle for those who embrace total monitoring of their lives, from food intake to daily activities—a hallucination of a perfectly optimized life that is. What lies behind the idea that AI may offer optimized solutions is the belief that one can foresee the effects of every action, forgetting perhaps Kierkegaard's insight that "life can only be lived forward and understood backward." Some would say that life is a sum of decisions. When we make decisions that influence others' lives as directly as they do in healthcare, we want to make sure we actually take the right decision. Automated systems using evidence-based AI algorithms promise an a priori *hindsight* in our decision-making, and that is why they are so seductive. They can also be misleading.

The users and beneficiaries of AI come from a variety of cultural spaces with different representations of health, body, privacy, well-being, and ethics. For example, patients and health providers may experience care in the same institutional context but in intellectually and emotionally different ways.

The simple interposition of a device for recording electronic medical data, for example, sometimes generates two contradictory feelings and attitudes. On the one hand, some patients feel neglected while the health worker is absorbed behind the screen, and for all the patient knows, while she or he is in pain, the health worker may as well be playing Solitaire. On the other hand, the health worker feels both pressed by the urgency of giving care and by the necessity of spending time in front of the computer, just an outstretched arm away from the patient, but completely absorbed and unable to provide comfort.

In even more dramatic ways, automation systems may be used for gathering information about, or even taking decisions about, the allocation of limited care unit resources in ways that may literally end lives. At the same time, the algorithms implicated in these decisions are more often than not inscrutable, not only for health workers but also for the IT specialists themselves—the famous metaphor of the algorithm as a black box. Despite this, they are programmed with principles in mind that may not be as universal as we like to believe but instead highly dependent on cultural biases. Those biases are not only about the decision-making itself that may refer to a very particular matter (e.g., to send a patient for further examination or not) but also about the assumptions regarding the larger material conditions of the lives of patients and those of the facilities in which the examination takes place.

Imag(in)ed Diagnosis

Image-based diagnosis is one healthcare domain in which AI using machine learning (ML) and deep learning (DL) for image recognition proved highly efficient. So high is the efficiency that many radiologists fear their profession is soon to disappear (Ford, 2015). However, despite the hype around image-based diagnosis using AI, up until now, the most efficient and accurate diagnosis process proved to be the one that combines human supervision with machine diagnosis, in other words, when a specialist double-checks the algorithm results. We are talking rather about a symbiotic relationship (Russell, 2019) and not an outright replacement. However, human-machine symbiosis does not always work as designed in the lab. The study of the use of DL in diagnosing retina damage done by Google in Thailand (Beede et al., 2020) is pioneering in the field of human-centered research and evaluation in healthcare. While human-centered and patient-centered approaches are becoming more and more the norm in healthcare related research, this study is perhaps the first that looks systematically at how algorithms, technology, and human expectations intersect the reality of caregiving in the industry. The study took place between 2018 and 2019 and a first article was published in 2020.

The article rightfully observes the disadvantage, and even the danger of the lack of formal requirement, in the adoption of automated systems in healthcare with human-centered observational evaluation. The research revealed both the necessity and the urgency for these types of studies and evaluations, in order to make sure that the implemented automated system works not as designed but as intended in relationship with the users and beneficiaries. In this case, the expectation of efficiency was met by the reality of real physical conditions when the system was implemented. In the workflow of diagnosing retina damage using DL, a key step is the taking of retina photos. This step is fulfilled by nurses, who later fed the images into the image recognition machine. The variable quality of light, the use or absence of pupil dilating solutions, and the impossibility to take more than two shots per patient (otherwise there was a risk of retina damage) sometimes produced images unusable by the machine but perfectly readable by a human interpreter. In many cases, two lower-quality shots from different angles could be read together by a human and give a complete image of the retina state. However, the machine only accepted one shot per patient. According to the workflow, in case of the system's refusal of the image, the patient should be automatically referred—even if a human could have read the image accurately. For the purpose of the study and for the implementation of the system, the decision to further refer the patient to a specialist had to be based solely on the machine reading.

Now back to humans, nurses were attuned to the patients' difficulties in their attempts to reach the specialists, which could imply a full day of travel using public transport or the necessity that the patient have their children who own cars drive them, disrupting their work schedule. In many cases, nurses considered that patients' comfort was an integral part of caregiving. Therefore, if the image was refused by the machine but readable by humans, they would "workaround" the workflow and choose to not further issue a referral for the patient. The results of this study point out similar issues and challenges to those described in the first chapter, regarding the Canadian clinical setting. The nested systems of care, algorithms, family ties, work relationships, interior design, and public transport act interdependently and create a new reality that is unforeseeable in the lab conditions in which any system is designed. Culture as embedded in both relationships and materiality, different understanding of what giving care means, and particular and particularly diverse arrangements in healthcare settings that do offer the same services speak to the necessity to understand the context, understanding that can be acquired only through observational, human-centered methods of research and evaluation. Adoption and adaptation of automated systems together with the hype created around them continue to reveal complex and rather problematic interactions between AI models and the reality on which they are deployed, as well as the conflict between competing AI ethical principles. Let us turn for the moment toward the latter and discuss it in the context of the COVID-19 pandemic.

Ethics: Principles Versus Practices

Former UN Secretary-General Kofi Annan said in his lecture at Tübingen University: "One thing that should be clear is that the validity of universal values does not depend on their being universally obeyed or applied. Ethical codes are always the expression of an ideal and an aspiration, a standard by which moral failings can be judged rather than a prescription for ensuring that they never occur" (Annan, 2003).

This is a powerful statement, touching at the heart of the ethical challenges of leadership. It may also contain a major flaw: while ethical codes can be framed as an expression of universal aspirations, the standards by which we could judge moral failings cannot equally be universal. Whether we like it or not, morality is culturally dependent—and, in many cases, moral failings fall into a cultural blind spot. This does not mean that, in the name of "cultural particularities," we can abdicate from universal ethical codes (although this is a current practice among authoritarian figures, particularly regarding the respect of human rights). It merely means that we need to be aware of how these aspirational universal codes are expressed in daily practices and how the transformation of these practices can (and does) generate new moral norms that, in their turn, shed light on those very cultural blind spots. Valuing human life is a universal ethical code. But what type of human life is more "valued" than others in different societies? And how do these societies make decisions based on that? Is a young life more valuable than an old one? How is this valuation expressed in daily practices? From whose perspective is the decision taken? There are cultures that are biased toward youth, while others favor a long-lived life. Is it more important to save the potential of a yet unlived life or the accumulated experience of a fully lived one? Is life at any cost more valuable than the individual choice to "not resuscitate" or dignity in dying? Is it possible to have an "equally valuable" approach to human life even in moments of scarcity? Is collective survival more important than individual well-being, and can they be separated after all?

These types of questions emerged forcefully during the pandemic crisis, and scientists, ethicists, and politicians continue to tackle the answers (or act as if they know them already). Two major conversations have dominated the public sphere lately: the use of data tracking systems for forecasting viral spread and the possible use of AI as decision support for the allocation of medical resources in emergency situations.

By looking at these conversations, we can see clearly how, sometimes, ethical codes may compete with one another. We will also catch sight of how the attempt to codify them in universally applicable AI algorithms can lead to the emergence of new types of biases instead of eliminating the existing ones. Thus, both deciding on the instances of using AI, and designing and relying on AI to make decisions need to have as starting points the practices that embody moral norms and not universal ethical codes and their presumed possible codification in AI algorithms. The immateriality of AI models received a reality check and so are about to experience the AI ethics field itself (Mantelero, 2022).

A Material World: Again

At a higher level of analysis, the pandemic in fact seems to be an eloquent reminder of the fact our world is material, despite a discourse of virtualization of everything, from markets to life itself. All of a sudden COVID-19 forced us to experience at least three major types of materializations:

Materialization of borders—while borders are not always easy to cross, and some are more material than others, during the COVID-19 pandemic years, the cross-borders movement had almost come to a stop. During 2020 and parts of 2021, most countries in the world had become inaccessible to others other than their citizens or residents. Supply chains had been disrupted, and their remaking is by far not similar with the prepandemic situation.

In parallel, extraordinary forms of collaboration at regional and global levels showed that only an open approach may offer long-term solutions, for example, the German hospitals uptaking French patients at frontier regions in order to relieve the overstretched French hospitals. The rapidity of developing vaccines was due mainly to intensive collaboration. At the same time, displays of solidarity were also suspiciously received, raising questions about using solidarity as mechanisms of soft power, particularly in China's case when the country decided to send medical aids to those in need.

(De-)Materialization of movement: Movement has become materialized and virtual, both at the same time. During lockdown, it has entered a controlled phase at all levels, with much of the workforce passing through a sort of mass experiment: work from home. Many of those for whom the ability to move was as natural as breathing

experienced it for the first time as a privilege. Movement was displaced to online platforms and dematerialized into bytes and pieces of data (more on this in a bit).

Materialization of our bodies—most importantly we are called upon to acknowledge the importance of our bodies. We, individually and collectively, have realized in a dramatic way that our lives are very real and unequivocally linked to our material bodies. The variations of the abstract indicators of the economy showed that the entire global complex system is not separated from but heavily dependent on our human bodies, their health, and their movement. This could contribute to the gradual dismantling of the illusion that we live in a virtual world in which the body is merely an instrument, among others, a tool to be refined in gyms and yoga sessions, or a resource to be overstretched during long, caffeine-fueled working hours. Somehow our bodies could become ourselves again. The reverse is also possible, that is, the desire of world-alienation may increase, facing the frailty that has become clear during the past few years.

Tracking

The data tracking systems (DTS) are not a novelty—the US police, for example, use it at large scale—nor is their use by marketing companies that rely on the individual users' data to push products through targeted advertising. The question of online data tracking was brought to the public's attention as early as 2012 (Madrigal, 2012), but the generalized use of smartphones made possible the extension of tracking from virtual movement to material movement in space and time. Apps that use the phone GPS (many times as a default option) track, store, and sell movement data to third parties for targeted marketing purposes. The same data can be used in instances by police in order to "prevent crime"—a practice that is both contested and not yet fully understood, let alone regulated (Friedersdorf, 2019).

The European Union (EU) enacted a data protection act that obliges developers to design, pseudonymization, and/or anonymization of users' data and to fully inform and to obtain consent from the consumers regarding their use of data (General Data Protection Regulation 2016/679, implemented starting May 2018, aka GDPR). This regulation impacts the use of data tracking systems and makes it more difficult to apply it indiscriminately or sell it to third parties (like US-based corporations tend to do). More recently the EU adopted a series of policy papers regarding the more general use of AI. With the COVID-19 pandemic, eyes turned toward the possibility to use DTS in order to predict and prevent the rapid spread of the virus by creating early warning mechanisms. The idea is relatively simple: once downloaded, DTS Apps track the user's movements using their phones and identify if the user is in the vicinity of someone who has already been registered as COVID-19 positive. The App would then alert the user and also create an anonymous map of possible viral spread.

AI ethicists raised first concerns, particularly regarding the tension between two AI major ethical principles: the user's autonomy (including rights to privacy) and usage for common good. First, the DTS cannot ensure 100% autonomy, especially when the GPS system is used for tracking. When movement data becomes health data (as in this case), anonymity is all the more important. Individual health data are highly sensitive. They are stored in highly secure environments, anonymized and used exclusively by specialists in healthcare. But what if the movement becomes a predictor for health? What if movement data are sold to third parties, interested in evaluating it through approximate models, in order to anticipate the individuals' health risks? Based on this data, can we decide who can and who cannot return to work, travel, or even visit friends? Or who receives scarce treatment if those who need it are too numerous?

Up to now, the ethical solution for the dilemma concerning movement data is a greater degree of anonymization. The design solutions proposed for the app are two-fold: use the Bluetooth systems of the device instead of the GPS to signal proximity only (and not location), and store all the relevant data on the device (not on third-party servers). Downloading and using the app is voluntary. A diversity of Apps featuring these solutions are being currently deployed. The USA had a fragmented approach; they granted the companies the decisions on the design and the deployment of tracing apps, as well as with the subsequent ethical decisions. European countries, on the other hand, have a more centralized approach, in that the government is more involved in financing and developing the apps, with features that need to meet the European standard of privacy. At the time of writing, Germany only started the deployment of their 20 millions Euro app and reassures its users that the data is not accessible to the platform provider they use (Android or Apple) but only to public healthcare specialists in the country. At the same time, Norway decided to withdraw its own app because the reliability was questionable, to say the least. Being based on voluntary download and reporting, and built on the assumption that people always carry their smartphones with them, the Norwegian government concluded that the App models do not necessarily correspond to human behavior.

For an external observer, the situation seems to be completely different in countries that appear to have a centralized, all powerful system of data tracking and AI use, such as China. While a European observer may conclude easily that the balance between common good versus individual anonymity is already tilted toward the first in China's case and that China already can use the Social Credit System in order to track and prevent COVID-19 spread, the reality is not precisely this. The approach in China seems to be more fragmented than in some European countries. Some provinces developed their own DTS, some of the apps use GPS while others are based on voluntary input of location by the user. Regional governments and cities may use different apps that may result in different "health scores" assigned to the same person. As Ding (2018) observes in his analysis of China's AI strategy, the Western perception is that AI deployment in China is top down and monolithic, hypercentralized, and controlled, with no room for ethics, far from the truth, as Ding shows in his work. In parentheses, this perception is a common trope of the depiction of "the East" in Western popular thinking. The doctrine of social peace and its attainment does guide the actions in China; however, ethical debates are present and led by private enterprises, such as the Tencent's Research Institute.

What DTSs actually do is combine borders, movements, and bodies and recreate them in the immaterial world of algorithms, while juxtaposing them to predesigned models, assigning the individual user to typologies within the models. It does not matter if the models are of consumer habits, potential delinquency, or the likelihood to pay one's mortgage. The trouble with the AI models is largely discussed in the specialty literature (O'Neil, 2017; Broussard, 2018; Galison, 2019), and it emerges from a few major sources:

1. Models are based on previous behavior and aggregated data and have a probability rate of correct identification. They are not 100% accurate! This may not be a major problem in cases in which we have models of success for an athlete's performance, but what happens when they are used to decide upon the finance or the justice delivery for individuals? Another drawback is that models function as long as the reality matches the conditions of the data collection, and they are highly dependent on the data quality and accuracy. Under normal circumstances (read long periods of status quo) the models more or less function as designed (emphasis on as designed). But as the COVID-19 crisis showed, a sudden disruption can cause "model drifting," that is, models no longer correspond to reality, and they need to be redesigned. This drifting was first signaled in the Amazon use of AI and then spotted in all major industries that use artificial intelligence (Castellanos, 2020; Heaven, 2020).
2. The use of proxy measurements in order to decide how values are attributed within a typology. For example, such a model may use the students' exams performance in order to determine the value of an educator, disregarding the whole series of other factors that impact on the exam results, apart from the educator's qualities and qualifications. At the same time, an educator who feels his livelihood threatened may be compelled to polish his students' results—giving birth to further distortions (see O'Neil, 2017).
3. Models are designed by humans and more often than not they embed the biases held by designers and developers. This is also a topic highly discussed in AI ethics. The solutions offered range from increasing the diversity in designer and product development teams to abandon the use of the tool itself altogether.

In conclusion, the use of DTS poses ethical dilemmas because they reveal the opposition between individual autonomy and common good. They also raise practical issues regarding accuracy and efficiency because of the way in which data is collected, stored, and used (Canca, 2020). DTS also raises an interesting issue about healthcare and data generation and usage: any data about a person can arguably be used as healthcare data or as marketing as convenient. Lifestyle can indicate stages of life and states of health. This raises the question of interchangeability of data. Individuals as data generators become interchangeable between the systems of knowledge interested in their lives, functioning on models that are by default just this: models imbued with cultural assumptions and personal biases. How are we to relate to the vast amount of data potentially available about an individual when a decision of life and death is made?

Back to Triage

The COVID-19 spread put an important strain on healthcare systems all over the world, and each country has approached the crisis in different ways. From avoiding testing and sending patients that were not in a critical state home—as it happened in the UK at the beginning of the pandemic—to carefully planning the lockdown and bed allocations in places like Germany, we have witnessed countless examples of uplifting solidarity action. The insufficient resources, such as hospital beds and respiratory units, and the necessity to allocate them to an increasing number of patients in critical states put a lot of pressure on the medical personnel. Ideally, all health systems should have guidelines for extreme situations such as pandemics, indicating how to triage the patient and allocate scarce resources. More often than not, the guidelines contain a set of recommendations about triaging the patients and allocating scarce resources, but they do not necessarily contain the practical ways in which the recommendations are to be implemented—so nurses and doctors are left scrambling to create their own procedures in this type of emergency.

The particularity of the pandemic put strain on the Intensive Care Units (ICUs) rather than on Emergency Rooms (ERs). The ERs around the world currently use a diversity of triage systems, and there usually one decides what type of treatment a patient needs and in what order of emergency. This is different from the pandemic situation in overstretched ICUs, where the treatment may not be available for everyone who needs it, and access to it has to be selective. This is an important distinction, and this is what happens today in many ICUs around the world. ER triage procedures do not apply to this situation. So what do healthcare providers around the world do? They are trying to follow the recommendations and to devise their own procedures in order not only to best serve patients and the common good but also to reduce their own enormous emotional stress. There are few criteria they may use, and as Philip Rosoff, ethicist and MD at Duke University, explained,[1] we know how not to take a decision of this kind: not in a rush, not at the bedside, and not using judgment based on privilege. In his words, in healthcare at least in the USA, there are ordinary situations in which there is a distinction made between VIPs and VUPs (Very Unfortunate Persons). In the case of the COVID-19 pandemic, this distinction is eliminated and so is the question of age. Age is not a decisive factor in providing treatment in case of scarcity (contrary to what an AI ethicist argued in a conversation on this very topic). The only criterion that should play a role, Rosoff affirmed, is the clinical chances of survival that the patient has. This can be assessed by healthcare professionals based on the healthcare records of the respective patient and on the current clinical state displayed. Here, one can see that artificial intelligence powered tools may come forcefully into play. Electronic Health Records (EHR) facilitate the preservation of patients' medical history and combined with the data of the current chart of a patient they could, theoretically, match the patient's

---

[1] Podcast with David Remnick https://www.newyorker.com/podcast/political-scene/in-a-nightmare-scenario-how-should-we-decide-who-gets-medical-care

history and current state with a recommendation regarding a triage decision. This may provide a certain relief in the high stress situations, and the decision may be supported by this type of evidence-based approach. However there are two important factors that need to be taken into consideration here:

1. The AI models' assessments of patients' health may be themselves flawed: the use of proxy measures in order to establish the health state of a patient (such as money spent in the past x years in health-related issues) can be very misleading: for example, such an AI-powered tool kept showing that Black patients' health is much better than White ones, and as a result they may receive less medical attention. The fact that less resources were spent on patients identified as black was considered by the AI system to be a sign of health. This was in fact due to the reversed type of causation: Blacks in the USA received less medical attention due to financial hardships and systemic racism, resulting in less money spent on health. If the conclusion is that the Black population is healthier, and a subsequent decision is to take this as a premise for further actions and policies, in fact it will continue the spiraling of inequality (Obermeyer et al., 2019; Himmelstein et al., 2023).
2. The risk of errors induced by the way in which the humans interact with the machines. One important aspect in AI as a decision support tool is that the system should remain a tool for support and not be transformed into the decision-maker. However, the high emotional stress amplified by the workload experienced by health workers may generate the so-called suspension of clinical thinking—that is taking the AI recommendation as the ultimate authoritative decision. In other words, under a variety of circumstances, particularly high stress, humans may be tempted to evacuate the weight of the decision to the machine. While this may be possible for a driverless car, it may prove disastrous in medical settings. Ironically, it seems (although it is not) easier to create an algorithm advising doctors (because everything happens between the screen and the health worker) than an integrated AI system that drives a car.

In conclusion, although AI may provide assistance in patient triage for resource allocation in a pandemic situation, it should not be transformed into an automated decision-making instrument, precisely because, as we have seen, biases and model dysfunctionalities may create irremediable medical errors. As with the case of retina scanning, the use of DL algorithms to decide on referrals, or in the case of ER medication administration, we need to develop workflows that integrate the human-machine interaction and avoid overreliance on the standardized answers of the machines. And of course, there is the question of accountability.

AI Ethics as Models

Both instances analyzed above (DTS and possible AI use in triage for medical resource allocation in ICU) have in common the question of ethics. When we try to make an ethical decision based on allegedly universal principles (such as "Life is the most valuable asset"), we should always be aware of the ethical codes that unavoidably vary from one context to another (between different social classes, working contexts, and so forth). What we practically (and many times unconsciously) do when we take a decision (or allow AI to take in an automated manner) is to hierarchize those principles or codes and to put them in a particular sequence. This is why an ethical decision taken in a certain context may look unethical if we apply it to another context.

According to ethical decision-making theory, there are five major methods to arrive at an ethical decision: the utilitarian approach (to make the most good and the least harm), the rights-based approach (to best protect the moral rights of those affected), the fairness and justice approach (to insure that the decision is fair), the common good approach, and the virtue approach (to decide the decision-maker values).

These methods are present and expressed as AI ethical codes in most of the approaches. Currently a series of organisms are stating principles for ethical AI that ought to be taken into consideration when designing and using AI. The EU put forth seven principles for trustworthy AI: human agency and oversight, technical robustness and safety, privacy and data governance, transparency, diversity, nondiscrimination and fairness, societal and environmental well-being, and accountability. Each of these principles comes with a list of recommendations meant to explain what they mean. Under privacy and data governance, we may find anonymity and respect for individual rights. Under societal and environmental well-being, we may find concerns for common good. As argued above, these principles may compete in different cases. They are also highly abstract, and they may be interpreted differently in different sociocultural contexts.

However, AI models interact with institutional, social, and cultural contexts and may fail if they are not designed for the context. This happens in fact in most of the cases in which AI works directly with humans. Ethical models do the same, so in order to avoid their drifting, we should develop them by firstly observing practices. The ethical codes themselves do not exist in theory, despite the fact that some ethicists generate them first theoretically. In fact, they are first expressed in different practices. Their very meaning is translated through practices, but practices vary in time and space. Different practices show the crack in models—as in the AI deployment cases. We should look at practices and their variations first in order to make our way back to judgments on values and ethics. Returning to the question of rights and valuing life—how is this expressed in various practices? How can we design decision-making mechanisms (automated or not) that correspond to the variability of practices and their dynamic transformations?

What Did the Pandemic Teach Us?

The major lesson for AI, and for ethics that COVID-19 brings, is that adoption means adaptation in a world in which matter matters. Therefore:

1. AI is a tool—and as such it cannot *be* ethical in itself, the ethics are always humans' business. Attributing ethics and bias to AI obscures human responsibility (Bartoletti, 2020). Instead it should be designed in accordance with ethical principles understood contextually, leading to humans acting ethically within the context of its use. Therefore, we first need to understand the context of its design and training, as well as that of its use.
2. Always set off having in mind that models are always wrong. Models do not drift because people behave weird. They drift because they are models. Their accuracy is limited in time, and the faster we change, the faster they drift. Their very introduction in use may be the root cause of the changing in behavior that in turn renders the model obsolete. Carrying models across cultural contexts will implicitly make them drift. That is why we first need to study the cultural context (regional, institutional, professional) and to work our way back from here into the design of AI systems.
3. The design process should start in the field not in labs. We need to design AI for specific cultural contexts, building models starting from reality, not trying to model reality according to abstract lab-designed models—and this applies to ethical models as well. Sooner or later, lab-made models will "drift," and ethics is one of the most important domains where they fail.
4. We need to create constant evaluation feedback loops. Remember, AI is material. It has a material support, and it interacts with the material world which means it is not going to flow smoothly. The environment will always trump its modeling.

The rippling effects of the pandemic are felt in economy, culture, and politics. In terms of AI, this means both a great opportunity to see where it is really helpful and a wakeup call to demystify some of the hype around it. One major lesson is that AI not only interacts with a material world in continuous transformation, but its functioning depends on this very materiality (and material culture). The crisis also reemphasizes the importance of understanding sociocultural variations—geographical or institutional—when approaching ethics. Moreover, it urges us to be more aware of the fact that AI design, deployment, and adoption has ethical implications. One major question that was overlooked till recently would be as follows: What domains and instances need the deployment of AI? Is decision-making based on AI appropriate for all the domains where we stubbornly try to implement it? Should we optimize everything just because we can? As Rosoff observed in his dialogue with David Remnick mentioned before, the US healthcare is a multibillion-dollar business, so in this particular context, the AI optimization may not always be in the best interest of the patient. Let us be patient and better ponder on where AI is really useful and where it has the potential of becoming a "weapon of math destruction" (O'Neil, ibid.).

The Bulldozer of Optimization

In 2019 at the Unfinished festival in Bucharest, I organized and mediated a conversation between Angelica Dass, a Madrid based artist born in Brazil, who is most famous for her project Humanae, and Moran Cerf, an Israeli neuroscientist now based in the USA, specializing in human-machine interaction at the brain level. He explores the ways in which technology may hack into consciousness. One of the questions I asked was what would a life perfectly optimized by technology look like. More specifically, I proposed to imagine how life would be if most of our decisions were to be taken by an algorithm that had access to all our data, from genetic makeup to neural mapping to every bit and byte of lifestyle choices made up to the present. What would they themselves do if the machines indicated a kale and salmon sandwich on oat crackers, perfect for their health improvement, over the butter croissant they crave? While Angelica chose the croissant, Moran argued that he would prefer to use his cognitive power for things more important than trifling choices and would go for the sandwich. He also hoped for a prolonged healthy, happy, and productive life. A day before, Moran had explained that the happiest moments for our brain are those in which we sleep. I could not help myself but thinking (and saying it out loud) that the hyperoptimized life he has in mind looks like a machine-guided sleepwalking.

So, what does optimization mean in this example? What is it that we optimize, and where do we leave space to suspend our decisions and follow the algorithmic ones? This is a fundamental ethical problem to be added to current AI debates.

Ethics may look like "universal laws" but, put into practice, they are decodified according to a series of very different societal, religious, legal hierarchies. We will need to deeply understand the effect of this fact in our rush to automatize. The hierarchies are furthermore interiorized by individuals into unconscious biases, and further down these biases are embedded in algorithms. Identifying this variance has been the focus of many mental and physical experiments, from the trolley dilemma to the moral machine (Awad et al., 2018)—the latter being essentially a gamification of the former. Proposed as a method to identify culturally specific norms by inviting people to answer a series of questions regarding choices made in different accident scenarios, the Moral Machine experiment offers results that classify different cultures based on the choices made by free participating people around the world. The result looks like a map of cultures on geographic and political lines (i.e., countries) based on the choices prevalent in the participants and claiming that it indicates ultimately "the types of lives" that people in a certain culture value more. This in itself is a problematic result, since it is based on both lab/gaming conditions and weighing the results but more so since it creates in its turn a model that, if used, forces itself on the dynamic reality of cultures. Should we build machines modeled on decision-making models based on modeling of real-life situations and then pretend that they are both rational and culturally sensitive? This is at least a triple-layered modeling here that removes the decision-making from the real-life situation and that models and standardizes ethics in a rather narrow set of assumptions. No easy answer is at hand. In this context, I think we should remove ourselves from the imperative of optimization, and the questions are as follows: How are we to optimize these types of decisions? Should we optimize them? And how precisely are we optimizing them?

This brings us back to the initial observation in ER, in which following indiscriminately the optimized workflow for medication in order to feed the automated system, the supposedly correct data in a standardized manner would jeopardize the caregiving process and even render it difficult to save lives. More often than not efficiency is defined independent of context and fails to account for the complexity of the process involved in any work environment. The view from the finance department, for example, of what efficiency means can be at times at odds with the perspective of quality assurance, and the way quality is understood in the latter may be completely different from what persons on the work floor strive for. Efficiency is not always effective. The third major question is brought to mind: *What do we optimize (for)?*

Brief Genealogy

Everything comes from somewhere. This includes speech (we all speak from our own subjectivity and experience) but also the way we make things and the way we conceive of them and their functioning. Objects tell stories; they embed ways of doing and ways of thinking that can be historically and culturally located. They carry in their design and functioning the visions of the world of their makers, while their users partially accept it and partially modify it, perpetually generating new ways of being and doing.

There is a fascinating tendency in our daily relation with material culture. Once we face an object, the ideas behind it are not only objectified but also perceived as *objective*. Once in use, an object seems to have the power to render permanent its own ephemeral and accidental, after all, embodiment. Because it functions, it creates the illusion of permanence for the ideas that it embodies. For example, once we step through a door, we tend to think that the door was always like this; *objectively* this is a door. It has always been; it will forever be. The very fact that it is solid, and it functions convinces us that it is eternal, and eternally this way, that it could not *objectively* be different. Subtle changes along the years are not registered as such. However, there are so many ways to organize the passage between two spaces; there are so many ways to conceive of doors across cultures and along history that we do not acknowledge. Does “the usual door” have a lock or not? Is the lock generally accessible with a key or heavily personified? Do doors slide, rotate, or revolve 90 degrees around their joints? From the light sliding Japanese doors that embed a completely different perception of space, relationship with intimacy, power, and control than that of the heavy insulating doors that became the norm in the standardization of buildings or than that of the absence of doors altogether in huts in some tropical or equatorial regions. However, they are all persuasive once we are within their universe of use. They contribute to the entrenching vision of the world they embody; they are part of its objectification and thus of its objectivity.

As a product of material culture, artificial intelligence is not an exception in this regard, and the ideas behind it are certainly "from somewhere." The very fact that it postulates rationality places it as if it comes "from nowhere," that is, from the vantage point of objectivity (only recently invented as we have seen). It presents itself as objective and thus universally valid. A closer look though will help us situate it in order to better understand what it represents at this moment of its history. So, what is the story of AI, and whose stories is it embedding, perpetuating, dispersing, and perhaps imposing with its persuasive power?

Artificial intelligence has a short, a medium, and a long history: the relatively short history of the making of electric-based automatic systems starting with the Second World War, the medium history that started with the transformations of thought starting with the Galilean revolution, the Reform and the Enlightenment period in Europe to which I alluded in the first chapter, and the long history of the dream of automaton entangled. These are the layered geological strata of AI and automation, and there is both continuity and ruptures between the layers.

Among many other objects of awe, the Hellenic civilization was known for its production of *automata*, mechanical devices moved by the power of wind, water, and Earth's gravity. They were used as functional devices such as a water dispenser actioned by coins and regulatory machines for politics (clepsydras that indicated the time allowed for a speech in the public space of debate (*the agora*), as educational devices, as wonder machines, or for artistic purposes. Some of the automata represented moving animals or humans engaged in diverse activities.

In the same period, the dispute around the functioning of the world, or more precisely what was the internal cause of movement, took place between the "atomists" or "mechanicists" and the "theologians." The former postulated that the constituents of the objects and their relationships offer the entire explanation for the movement and animation. In this vision, beings and objects are both governed by the principles of mechanics physics, and the mechanical explanation suffices to explain the functioning of the universe and everything in it. If only we knew precisely the materials and the way they mechanically interact, we would know the behavior of any object in the universe, alive or not. Here one can identify the roots of philosophical materialism, a restrictive view of the world as Silvya Berryman calls it, in that it restricts explanation to the components and their interaction governed by mechanical laws.[2]

The teleological view, on the other hand, tells a different story: the naturally animated objects such as animals and humans may be imitated by mechanical machines to a certain degree, but they are animated by a teleological principle that adds to the pure mechanical explanation. Aristotle calls this "pneuma," the principle that is beyond the materialistic explanation of an animal behavior. In this view, for

[2] Ancient Automata and Mechanical Explanation Author(s): Sylvia Berryman Source: Phronesis, 2003, Vol. 48, No. 4 (2003), pp. 344–369 Published by: Brill.

example, animals and humans can be imitated by automata only to a certain degree but not completely, because the automata fundamentally lack the teleological principle. The automata may give the impression of self-generated movement, with a *thelos*, if the mechanisms are hidden from view, as Hero of Alexandria, an automata designer and constructor, points out. This is the only way, in his view, that the sense of "magic" can be conveyed. One has to render the mechanism small enough to be invisible to the viewer, so that the automaton will *look like* the real thing for the spectator. However, it can never be equivalent—obviously lacking the *pneuma* in Aristotelian terms.

These two competing stories about the universe and its functioning are carried over in the history of European thought, each of them dominating one era or another. The rise of early Christianism and the dominance of monotheistic thought declared the teleological vision, with its variations, a winner for the next ten centuries. The *pneuma* became the soul that which is eternal and subject to the will of God. Through the vehicle of the soul, the will of God manifested in the world. A new hierarchy of beings emerged, in which plants fell to the third place, being apparently inanimate, animals were above them, and humans on top, endowed not only with souls but also with reason. The fall from grace—in the biblical story that became predominant in Europe and more broadly in the societies influenced by Abrahamic religions—endows humans with reason and consciousness of the sin. Sin and consciousness are thus intimately entangled in this relatively new vision of the world: awareness of one's sinful nature places one above the animal kingdom. However, animals hung on to their souls for the next centuries, until the mechanistic view returned under a new guise: modernity. The entangled transformations discussed in Chap. 1, scientific, geographical, and religious, led to the elevation of the human soul to an affair of divinity solely, the placing of pure reason as the only valid and acceptable way of being in the (profane) world, and the loss of soul for the animal kingdom.

In fact, the disappearance of the soul from earthly affairs created a place for the newly discovered Newtonian mechanical physics to enter gradually, metaphorically, or not, in every aspect of life. The world, including animals, became subjects of reason and objects for observation and their relationships explainable in the same way in which one explains any physics phenomena. Algebra and mathematical analysis, imported in the Renaissance along with the concept of 0 and the "Arab" numerals from, well, the Arab mathematicians, helped in constructing the argument of a perfectly rational and computable universe (more on this in a following chapter). Among the Arab thinkers important for our conversation, forgotten as a person but whose name is on everyone's lips when speaking about AI is al-Khwarizmi. Active during the ninth century during the Abbasid period, he was a polymath who advanced a model of reasoning in successive steps. This model advanced a method of following clear steps from logical inference to logical inference. This is not a model based on mathematics—the exact science of numbers—but on reasoning with propositions that have both a syntax and a meaning (semantic). The meanings are field dependent—and al-Khwarizmi was conscious about this. He became known in Renaissance Italy as Algoritmi.

Time gained the linear dimension of movement, and so did history, creating thus the idea of progress. This idea was sustained by the discovery of "deep time" along with the exploration of the depths of the Earth and the apparition of Geology marked by the publishing of James Hutton's book *Theory of the Earth* in 1795. All of a sudden, the Earth appeared to be not anymore young when humans appeared as the result of creation but much older than the human's blink of existence. The geological strata bore witness to this (Gould, 1987).

The concepts of mechanical physics became explanatory for human phenomena too and helped with the creation of the new concept of the individual as a monad, a mobile who/that moves according to one's (reasoned) internal will, much like a Newton's mobile.

Two figures of human automata can be contrasted in order to perfectly illustrate the way these opposing conceptions became embodied in monstrous metaphors: the Golem and Frankenstein's creation. The difference between the Golem and Frankenstein's monster emerges in the very ways in which they become animated: the former is brought alive with incantation of mystic words passed from generation to generation to the initiated ones and the latter by the activation of the matter with electrical impulses—electricity itself being a product of materialistic view embedded in physics. The former is an expression of a teleological vision, with a simulacra of a soul, a ghost of its maker's soul, while the latter was a material creation with scientific methods whose functioning can be entirely explained by the description of its parts. While the Golem is teleological, albeit with a human creator, Frankenstein's monster is by definition the expression of enlightenment and modernity's dream of a mechanistic Universe.

The nascent and soon to become dominant philosophy was expressed in Descartes' writings. Gradually the world became objective and the knowing subject, the human, separated from it. The illusion that the world could be acted upon without consequences on humans (because the human's existence is concentrated and expressed solely in thinking) gradually became predominant. It is an illusion in which we still live today. The industrial revolution and the mechanization of productive activities gave substance to this worldview. Its truth value was reinforced by the fact that "it works." And it does, indeed, at the scale of each of the fragments it acts upon. However, for a long while the image of the whole system is lost, and each of the fragments becomes an exploitable resource.

The first programmed machine in this economy is the punch card loom, which functioned pretty much like the early computers. Inspired by it, in the nineteenth century, Babbage, an English mathematician, imagined the first complex calculus machines, and Ada Lovelace the first modality to "talk" to them. She is considered the first programmer.

On the other side of the Atlantic, the birth of the newly self-proclaimed American nation was intertwined with conquest by means of mechanical power—with the train as its emblem—rapid industrialization and extraction of the resources from the "new and *empty*" land. The land itself was the most important resource and it was mathematically conceived using the grid pattern—a blanket of squares projected on the map with disregard to the characteristics on the ground, and based on which property, and power, was projected and distributed from coast to coast. The geography with all its characteristics was buried under a double layer of abstraction: the map and the grid on the map. Actions on the model (the selling of the land, for example, or the tracing of railroad routes) were afterward enacted by force, violence if need-be, on the ground realities. The would-be pioneer colonists could thus buy land from the map, and through this mathematical-topographic and political fiction, they could travel there and "rightfully" claim it. It mattered little what was there previously, because the grid on the map equalized reality and subsumed it to "wilderness." Of course the continent was neither new, nor empty, but the compartmentalized view of the world could put its habitants aside, separate them from the landscape of "wilderness" there to be tamed, and eliminate those who resisted the logic—simply because it would be irrational to resist progress. This literally happened with the genocide of indigenous people and the creation of reservations for the surviving natives.

The colonial expansion powered by mathematics needed more and more complex forms of calculations. The nineteenth century saw a growing demand for human calculators: people whose jobs were solely to make complicated mathematical operations in order to generate tables of complex numbers to be used by engineers and scientists in their respective endeavors. The solution sought for was obviously a technical one: time and financial resources were invested in the development of calculating machines. Obviously, the technological solution worked, and it potentiated the later development of computers. Less obviously is the fact that this would not have been the only choice. As Foer (2017) observes in his book, the human resources shortage was also linked to the fact that a great part of the population was regarded as not fit to work in these kinds of jobs, that is, women and slaves. Instead of using the path of social reform (slavery abolition and giving access to women in the workforce), the technocratic solution was pursued. I am not suggesting that efforts in creating mechanical calculators should have been stopped, but that this choice reflects a certain worldview generally about the hierarchization of work, and particularly regarding the STEM domain that is perpetuated to this day. It also points out to the nascent belief that to social ailments we can and *should* find technocratic answers.

Silicon Valley is at the end of the pioneer's road: California. The end of the road did not mean the end of the pioneering culture and ambitions. The dream of conquest and remodeling of the world to fit the technocratic view expands to both space and bodies. On July 10th, 2021, at the launch of his space rocket that propelled him out of the atmosphere for 10 s, Jeff Bezos donned a "ten-gallon" hat and belt over the space suit: a true space cowboy. Elon Musk is actively dreaming and planning the colonization of Mars, while wreaking havoc in a society that does not function (yet) based on his libertarian principles—and, well, does not behave "rationally" and is not "free." However, the freedom he and the inheritors of the pioneering culture envisage is one in which responsibility toward the other is not placed in high ranking. In Viktor Frankl's terms, freedom without responsibility is in fact arbitrariness and imposition of individual will based on power. Where can this be achieved?

In a place in which care for "the other" does not interfere. Space is one of them and the internet, another.

Silicon Valley's culture and the genealogy of a post-World War II approach on artificial intelligence is described in detail by Meredith Broussard in her *Artificial Unintelligence*. From Minsky to Steve Jobs, Elon Musk, or Zuckerberg, passing by Kurzweil and arriving at Thiel, Broussard shows the combined effect of the intersection of multiple visions that converge on the Cartesian fantasy of disembodied rationality. Science fiction dreams and the active effort to turn them into reality meet the conviction that technology could offer a solution to create a utopian world without conflict but also a world of absolute freedom, built by, and for men with disdain for social conventions—and for those different from them. This utopia was powered by vast amounts of free money pumped without any checks and balances all the way into the mid-1980s, mostly in MIT's Media Lab led by Minsky. A dose of nonconformism is always good; it makes things change. Nonconformism with nothing to replace it except the logic of financial gain has a different name: anarchocapitalism. "Move fast and break things" is its perfect motto.

As Hannah Arendt (2018, [1958]) pointed out, solutions are to be found in escapism from the confinement of the Earth as it exists now: in space or in technocratic fiction. But this fiction should be populated only by those who believe in it. Thiel explicitly makes the case for such a space in his essay *Cato Unbound* dated in 2009: "Since 1920, the vast increase in welfare beneficiaries and the extension of the franchise to women—two constituencies that are notoriously tough on libertarians—have rendered the notion of 'capitalist democracy' into an oxymoron" (quoted in Broussard, 2018: 83). At the end of her analysis of the people who created the AI, Broussard asks herself what can go wrong when you have "an elite group of men who tend to overestimate their mathematical abilities, who have systematically excluded women and people of colour in favour of machines for centuries [...] who don't believe that social norms and rules apply to them, who have unused piles of governmental money sitting around, and who have adopted the ideological rhetoric of far right libertarian anarcho-capitalism" (Ibid.: 85) deciding about the future based on their fantasy.

The answer is rather frightening: a world of arbitrary freedom absent responsibility, ruled by algorithms based on mathematical models that approximate reality but eliminate the variations that do not fit the model sound pretty similar to the grid pattern applied to a map and then enacted on the world. To keep with the metaphor, the map, the first layer, is the algorithm—an approximation that contains conventions and the worldview of its maker embedded in it. The grid pattern is the mathematical model that further fragments the map in squares presumed to be equivalent to each other. An undifferentiated world, a reproduction of the same ad infinitum, but to what end?

The genie is out of the bottle. Putting it back is impossible. Despite the gloomy picture painted so far, this question needs to be continuously asked, at every step. One of the frequently given answers is "optimization." In a repeat of Broussard's question: "what can go wrong if we optimize"? As an end to this chapter, I attempt some answers.

Fine-Tuning Optimization

At every step of life, humans face the need to make decisions. The "natural" desire to make the right decision meets the impossibility to know beforehand the effects of the decision. The decision itself changes the parameters of the environment in which the decision is implemented. If only we could know better, or more, or perhaps everything, when making a decision. We all want hindsight with our foresight!

For many, data-based technology and decision-making promise precisely this: a universal a priori hindsight, a complete control of the environment, and an almost perfect predictability of results—if only we had enough data. The automation system also promises rapidity in decision-making and optimization. From urbanization to decisions enhanced by the paradigm of smart cities, from data-based decision-making in healthcare to shopping lists and eating habits, etc., optimization permeates all aspects of life.

The desperate race to reach a desired objective in the shortest time and with the minimum expenditure of resources, while ignoring the way to get there, deludes us into optimizing the things that matter the least for our well-being. How can we address this almost natural tendency? Do we need a new type of thinking to navigate the pitfalls of automation?

What resurfaces here is not what data-based systems can and cannot do but how their users relate to them, what they expect from automation, and how optimization and optimal decision-making have transformed political decisions. Most of the time, the expectations are that the automated system will give us the best solution in the shortest time. But do we formulate the demand correctly? The classic example of failure here is the thought experiment in which an AI destroys civilization when programmed to make metallic paper clips. Given absolute power and access, the AI-powered machine would use every scrap of metal available on earth, the notion of what efficiency means being translated into making paper clips at all costs.

This happens on a small scale in many real cases when, for example, we measure efficiency in work environments. Efficiency is defined independent of context and fails to take into account the complexity of the process involved in work. For instance, let us think of the distribution of medication in a hospital. A measure of efficiency of work is the rate of medication scanning per distributed dose of medication in each department. This implies that the nurse would use a barcode reader to first scan the medication label, check the patient in the system, and then scan the patient's barcode on the bracelet attached to their right hand on admission. This may work well when patients are already hospitalized, but what happens in emergency departments? They may have a very low rate of scanning, thus failing lamentably in the eyes of the person checking the screen for the rate of medication scanning. Why? Because, for most health workers, saving lives by administering doses of medication immediately takes priority over the burdensome task of scanning the barcode on the patient's bracelet (if the patient has one in the first place). So what is more efficient? Saving a life or scanning the medication with the possible price of that very life?

Extreme optimization, short termism, and efficiency narrowly defined in monetary terms are pitfalls of automation that need to be carefully considered. In this context, it becomes obvious that a new type of thinking is needed, one that could perhaps understand and relate to the mechanisms of automated decision-making in a critical way, without people being seduced by its apparent magic.

On a practical level, it is perhaps advisable that we spend more time thinking about these three questions before implementing automated decision-making systems, before designing algorithms, and before considering the parameters of that design:

1. Why do we optimize? In other words, what does optimization mean? Reducing costs or increasing well-being and happiness? Answers to these questions may be found through better care given to the definitions of metrics and measurements when deciding on the usefulness of an automated system.
2. How do we optimize? Did we take into account as many variables as possible, including the sociocultural context and the type of activity we'd like to optimize? It is possible, probable, and even desirable to explore the array of unforeseen effects before designing and implementing an automated system in context?
3. What do we optimize (for)? Perhaps there are domains in which a certain type of optimization should be kept at bay or should be rethought for the domain itself.

Answers to these questions can be found only through observational, human-centered research and evaluation techniques that integrate ethnographic methods with institutional analysis, and design thinking. After understanding the context, we should approach optimization as a tool for a well-defined scope and not as the scope itself while usually using a very narrowly defined optimization definition. More often than not, in the case of AI algorithms and automated systems, the latter seems to be the case. Perhaps it is high time to ponder on the variance of optimization and AI in function of context by redefining optimization in the first place. And perhaps not every domain of life needs to be optimized in an algorithmic manner.

References

Annan, K. (2003). *Do we still have universal values?* (speech) Lecture at Tübingen University, Germany. https://press.un.org/en/2003/sgsm9076.doc.htm

Arendt, H. (2018 [1958]). *The human condition*. The University of Chicago Press. (Note the bracketed original publication date for a reprint).

Awad, E., Dsouza, S., Kim, R., Schulz, J., Henrich, J., Shariff, A., Bonnefon, J.-F., & Rahwan, I. (2018). The Moral Machine experiment. *Nature, 563*(7729), 59–64. https://doi.org/10.1038/s41586-018-0637-6

Bartoletti, I. (2020). *An artificial revolution. On power, politics, and AI*. The Indigo Press.

Beede, E., Wilcox, L., Baylor, E., Hersch, F., Iurchenko, A., Ruamviboonsuk, P., & Vardoulakis, L. M. (2020, January). A human-centered evaluation of a deep learning system deployed in clinics for the detection of diabetic retinopathy (study – conference paper). Conference: ACM CHI – Honolulu, HI.

Berryman, S. (2003, August). Ancient automata and mechanical explanation (article). *Phronesis – A journal for Ancient Philosophy, 48*(4). Brill.

Bridle, J. (2022). *Ways of being: Animals, plants, machines: The search for a planetary intelligence* (1st ed.). Farrar, Straus and Giroux.

Broussard, M. (2018). *Artificial unintelligence: How computers misunderstand the world* [Website]. MIT Press. Retrieved from https://mitpress.mit.edu/books/artificial-unintelligence

Castellanos, S. (2020). Pandemic-related disruptions hurt reliability of some AI models [Website]. *Wall Street Journal.* Retrieved December 8, 2020, from [invalid URL removed] some-ai-models-11591781402

Çanca, C. (2020). *Why 'mandatory privacy-preserving digital contact tracing' is the ethical measure against COVID-19* [Website]. Medium. Retrieved August 4, 2020, from https://medium.com@cansucanca/why-mandatory-privacy-preserving-digital-contact-tracing-is-the-ethical-measure-against-covid-19-a0d143b7c3b6

Ding, J. (2018). Deciphering China's AI dream – The context, components, capabilities, and consequences of China's strategy to lead the world in AI (analysis). Future of Humanity Institute – University of Oxford. https://www.fhi.ox.ac.uk/wp-content/uploads/Deciphering_Chinas_AI-Dream.pdf. github.com

Foer, F. (2017). *World without mind: The existential threat of big tech* (1st ed.). Penguin Press.

Ford, M. (2015). *Rise of Robots. Technology and the threat of a Jobless Future*. Basic Books.

Friedersdorf, C. (2016, January 27). *An unprecedented threat to privacy* (article). The Atlantic. https://www.theatlantic.com/politics/archive/2016/01/vigilant-solutions-surveillance/427047/). Accessed 4th May 2019

Galison, P. (2019). Algorists dream of objectivity. In J. Brockman (Ed.), *Possible minds. 25 ways of looking at AI* (pp. 231–240). Penguin Press.

Gould, S. J. (1987). *Time's arrow, time's cycle. Myth and metaphorin the discovery of geological time* (reprint edition). Harvard University Press.

Heaven. (2020). *Our weird behavior during the pandemic is messing with AI models* [Website]. MIT Technology Review. Retrieved from [invalid URL removed].

Himmelstein, G., Ceasar, J. N., & Himmelstein, K. E. (2023). Hospitals that serve many Black patients have lower revenues and profits: Structural racism in hospital financing. *Journal of General Internal Medicine, 38*(4), 586–591. https://doi.org/10.1007/s11606-022-07562-w

Hutton, J. (1795). *Theory of the Earth, with proofs and illustrations* (1st ed.). Cadell and Davies.

Madrigal, A. C. (2012, February 29). I'm being followed: How Google—and 104 other companies are tracking me on the web [Article]. *The Atlantic.* https://www.theatlantic.com/technology/archive/2012/02/im-being-followed-how-google-151-and-104-other-companies-151-are-tracking-me-on-the-web/253758/

Mantelero, A. (2022). *Beyond data. Human rights, ethical and social impact assessment in AI.* Springer.

O'Neil, C. (2017). *Weapons of math destruction: How big data increases inequality and threatens democracy*. Broadway Books.

Obermeyer, Z., Powers, B., Vogeli, C., & Mullainathan, S. (2019). Dissecting racial bias in an algorithm used to manage the health of populations. *Science, 366*(6464), 447–453. https://doi.org/10.1126/science.aax2342

Russell, S. (2019). *Human compatible: Artificial intelligence and the problem of control* (1st ed.). Penguin Books.

Part II
Past Traces of the Future

Chapter 4
Knowing Nature

What speaks to the soul, escapes our measurements
Alexander von Humboldt, 2012

AI apps hold not only the promise of empowering us with complete supremacy over our bodies but also with utter control of nature itself. We want to believe that AI can somehow give us mastery over nature by simply measuring and managing it. But how did we come to think that nature is separated from us, measurable, and manageable? And how is this conviction reflected in the way we generate knowledge about and understand nature today? What are we consciously or unconsciously leaving aside so we can build our predictable models that regularly fail?

Snake Stories

When I was a child, my grandma was enchanting our hours spent together with stories about the world surrounding us. I remember one time the two of us, watching the summer rain clouds gathering in the sky. She told me: "You know, the rain is brought by a dragon (rom. *balaur*) that gathers the waters in the sky. The dragon is a snake that goes into the sky, and every year it is a different one. The snakes gather together in the spring, when snow melts and the sun is warm enough. They dance together and form a ball, and you cannot see anymore whose head and whose tail there is, that's how many they are. And from that ball, a foam is forming, and the foam becomes a pearl. And they all try to grab the pearl with their mouths. Whichever catches it, becomes the rain dragon of that year, is pulled up in the sky and traverses it bringing us the rain." Many years later, in 2014, I saw for the first time a Chinese New Year procession in Vancouver. And, lo and behold!, the dragon with the pearl in its mouth was there, happily dancing on the streets, celebrating the rebirth of nature, calling for, and announcing therefore the fertility of future rains. At the time

A. Balasescu, *Climate Change in the Age of Artificial Intelligence*, Artificial Intelligence, Simulation and Society,
https://doi.org/10.1007/978-3-031-90042-6_4

of this writing, 2023, the ball of snakes appeared again in a story of a student from Manitoba, Canada. She told me in a conversation that there is a tourist attraction site in Manitoba where people go in the spring to watch the spectacles of the snakes coming out to congregate and copulate, producing precisely the images evoked by my grandmother. However, from her story, the mythology behind was absent. The phenomenon was retold and represented as the spectacle of nature revealing itself for consumption to the avid eyes of the tourists and occasional passersby. There is no expectation of the materialization of a pearl, let alone of the metamorphosis of a snake into a dragon ascending to the skies. Of course, there is no surprise in the disenchantment of the world in the early 2020s. But what are the roots and the effects of this disenchantment on our relationship with the environment?

The snake is a powerful symbol, ubiquitous in myths and mythologies of nature. In ancient Egypt, Uto, the goddess with a snake appearance was the protector of the kingdom, and the cobra was the crown on Ra's head or depicted near Horus's wings. Ancient Helenic culture had the snake appear as a single helix on Aesclepios' walking rod, a symbol of healing still present in some pharmaceutical logos (Fig. 4.1). In the same culture, we encounter the Caduceus, two snakes intertwined in a double helix, around a staff with wings at the top. Hermes, the god of trade, wealth, luck, fertility, and animal husbandry, among others, used to carry it.

King Naga, the mythical snake in Hinduism, curled his body around the Mountain Mandara that sits on the back of Vishnu as a giant turtle. On each side of Naga, Devas (demi-gods) and Asuras (demons) are pulling on his body and churn the Ocean's water for thousands of years to obtain Amrita, the elixir of eternal life. This

Fig. 4.1 Aesclepios on a building in Bucharest—photo by the author

Fig. 4.2 (**a**) The making of amrita, Angkor Wat, Cambodia. (**b**) The making of amrita, Bangkok International Airport—photo by the author

can be read in many ways, including as the myth of creation of life itself out of the constant confrontation of antithetical forces, but without the idea that there is a final synthesis. There is a continuous dialogue (Fig. 4.2a, b).

Fig. 4.3 Buddha in meditation pose protected by Naga—photo by the author at the Angkor National Museum, Siem Riep, Cambodia

In Buddhism, Naga appears in a very important moment in Buddha's life, when he meditates and a strong storm gathers. Naga, having seven heads, opens them up like a hood around and above Buddha. This protects him from the wind and rains and yet allows him to continue to be in nature, while integrating his presence without separation (Fig. 4.3).

The scene reminds me of the episode recounted by Jessica Hernandez (2022) in her book *Fresh Banana Leaves* but with a twist. Hernandez is a biologist, a professor at University of Washington, Seattle, and of indigenous origins. Her father survived a bombing attack during the Salvadorian civil war at the base of a Banana tree, protected by its leaves. The author uses this story to illustrate a different relation with nature: a protective nature that saves her father from human violence. Hernandez uses the story in contrast with the contradictory attitudes we display in relation to nature in modernity: a resource of both materials and also of "recreation," or a threat, but never an ally, a remote entity, not an embracing, integrating one as it appears in the story of her father.

In the *Old Testament* biblical stories, snakes are ambivalent, sometimes in the same parable. For example, when the people of Israel revolt after the crossing of the desert, God sends a rain of snakes and frogs to punish them, but he also instructs Moses to make a brass snake that provides healing to those who look upon it. The preexistent polytheistic gods appear in the Bible under the form of (false) prophets or kings, for example, King Baal, the god of rain and fertility whose wife is represented as a snake.

This ambivalence is discarded once and for all with the story of the exile from the Garden of Eden. That snake informs in great measure both how most of the people influenced by the monotheistic Judeo-Christian culture feel about snakes, and about nature. The fall from grace story is the story of separation from the natural state, again not in Rousseau's sense, but in the sense of the loss of connection with the nonhuman world, of the feeling, and conceptualizing ourselves, as being an integrative part of it. The snake offers the fruit of knowledge, but that knowledge, as it were, is a knowledge that *separates* self from the rest, the prefiguration of what was to come a few thousand years after the writing of the *Old Testament*. With the knowledge of creation came the consciousness of death—and the angst that accompanies it. And while death is part of creation (and of materiality/nature), the desire to escape it moves one to the illusory path of separation from that very creation. It is the first time in which the snake appears as a negative character, but I think that the fear it instills is not of the snake itself but of what comes after the knowledge of creation and death is gained and the desire of separation from nature begins. It is perhaps important here to mention that the word Satan, the absolute evil for which the snake is a metaphor, comes from the verb root *s-t-n* meaning exactly this: to separate and to divide.

The *Old Testament* itself marks in many metaphors and stories the passage from nomadic and pastoralist life to sedentary agrarian societies. The latter are the first ones that turn nature into a controllable and exploitable resource and dismantle the nomad organization, provoking its gradual disappearance, and also the wrath of God. The fable of Cain and Abel is this story, and we can imagine that the saddened and hurt God is Gaia herself, the whole planet system as it was theorized by Lovelock, capable of self-generating life, the quality known as *autopoesis*. She "naturally" loves the seminomad pastor Abel more than she does Cain, who reduces her capacity to the necessity of our species only. The jealous Cain kills Abel. After the original crime, the task she gives to Cain contains both the choice and the possibility of being in the world: "You can master the sin"—the sin being here the destruction of Cain's brother, that is, the destruction of a lifestyle more attuned to nature as autopoetic divinity.

Mastering

Thinking of where we find ourselves now, we did not do a great job in controlling the control-over-nature compulsions. Here, in these two stories, the fall from grace and the Cain and Abel fable, we have the germs of separation from nature proper to sedentary populations. What followed is the gradual refinement of the means of extraction of resources, mastering of nature, and completing of this separation by means of technology, from the plough to the current AI tools. There is perhaps an interesting observation to make, beyond any detailed analysis of tools and material culture that are the object of intersecting fields of anthropology, phenomenology, and philosophy. Tools coming from predominantly agricultural societies are made on a "catch-all" ideal. The plough is made to cover and transform the entire surface of the land. Fences—another sedentary tool—are made to clearly separate and exclude, and they are designed with different degrees of porosity, but they tend to become more and more exclusive. The modern deep sea fishing trawlers are there to literally catch everything in their way, transforming the bottom of the sea into something similar to an arid agricultural land and so on to the data-producing tools of our modern technology designed to catch all of the data-productive possibilities. The indigenous populations today still preserve tools that are not designed like this. One of the examples amply discussed by Mueller (2017) in his volume *Being Salmon, Being Human* are the fish baskets used by indigenous populations in North America to catch salmon during their migration. By design, the fish baskets are made to let pass through different sizes of fish. The underlying principle is that of sharing: sharing the possibility of continuing to live with the salmon population and sharing the food gifted by salmon's migration with all the other beneficiaries—from bears and bald eagles to trees and vegetation (more on this, later).

The expansion of sedentarism kept transforming the land from a giver of life to biosphere, to a giver of food to a very reduced part of the regnum, the humans and their domesticated lot. The scientific revolution discussed before refined the tools of exploitation. The nonconforming populations became more and more rare; Cain kept killing Abel. The culmination of this process is of course the colonial advent in which the meeting of the worlds generated massive destruction among indigenous nomadic and seminomadic populations. The lands encountered in the colonial advent were presented as belonging to the wilderness and their inhabitants as "savages." This set the scene for the "domestication" of both: the taming of the wild and the civilizing of the irreducible other, "the savage." But these categories were not preexistent; they were produced by this very encounter, as oppositional categories with the civilized, progress-oriented, and morally good "us." This latter was itself produced by the invention of the former. The morality of "us" was extended to the nonhuman realm, as "natural laws," imagined by the philosophers of the Enlightenment, and presumably governing human activity and the nonhuman regnum. However, this nature was a moral one, a reflection of the morality governing the times, used in turn to legitimize them. We continue to follow this illusion of a moral nature today; how many times does one hear that something as arbitrary as all-encompassing individual competition is "natural"? Not to talk about the fact that it is "natural" to have two, and only two, sexes. Everything falling out of this moral nature is assigned to wilderness, and therefore it is morally justified to take it into possession and tame it. If not possible, destroy it, humans included. It mattered little if the taming process itself implied destruction. Here lies the difference between Rousseau's "noble savage" and the indigenous populations met, and systematically wiped out, in the colonial encounter: the "noble savage" is morally good; the indigenous are "immoral savages." This is what Thomas King, a native novelist and scholar from Canada, calls the dead Indian as opposed to the live Indian who is simply inconvenient (King, 2008).

Those were not only destructive times but also productive ones. On the one hand, they produced these new categories of oppositional thinking: the civilized "us" and the savage "other," progress as opposed to backwardness and the scientific view as opposed to the supposed magical one. Of course this is a very simplified, schematic view of the events taking place during the colonial expansion. On the other hand, the encounters produced the very opposition to these binaries, and they also gave birth to new concepts to which we are tributary today, for the good or the bad. Anthropologists, starting with Malinowski, set out to dismantle this set of binaries, showing that the "savages"' methods for observing the world and acting upon it were akin to any scientific rules and methods, albeit not using the same standard measures. However, they were not used for the same scope that of the total domination of the surroundings.

The matters get even more complicated if we allow for the particularities of each encounter and to the great intellectual influence that the meeting with different forms of social organization had on European concepts of society and politics. In *The Dawn of Everything*, David Graeber and David Wengrow (2021) argue that keystone terms and meanings for modern democracy, such as equity, social equality, and political consensus, were borrowed by English and French thinkers directly from their indigenous counterparts. Even more telling, those who presented these ideas in the metropole giving credit to the indigenous personalities that inspired them were derided. In the view of their contemporaries, the "savages" could not possibly be so insightful.

During that time, the "view of nature" itself was refined to become the construct we know as nature today. It produced and generated most of the concepts with which the modernity-influenced world approaches the nonhuman world, notably that of the environment. Nature as we know it today was invented, along with the means of knowing it in current terms. One person of great power of observation, sensibility, and intellect had a particularly great influence on this: Alexander von Humboldt.

In the early nineteenth century, when Europe was shaken by the aftermath of the French revolution and the rise of Napoleon and his wars, Alexander von Humboldt set out to South America in a desire to observe and understand the continents across the Atlantic. He found there a world that was also in movement, both politically—the independence struggle led by Bolivar was under way—and from the perspective of the environment, the effects of the colonial presence on the geographies of the continent being already visible for his keen eyes. The result of his voyage is multifaceted. First he produced a vast body of writings that combined observations, measurements, descriptions, and analysis of the surroundings that were not limited to the natural world but included the political and social transformations that were under way in that period. Second, he shifted the way in which nature was perceived and politicized at the time, entering into direct conflict with the ideas of naturalists of the period and inspiring the generations to come. Third, von Humboldt established the importance and validity of scientific measurements, with effects that in fact he could not have foreseen and that were ultimately obscured until recently numerous insights of his work.

Alexander von Humboldt can be considered the precursor of Gaia theory; in his work, he emphasized the deep interrelations of all elements of the planet, humans included. And because he was a sensitive observer, he also understood the destructive character of the colonial advent, on both humans and the environment. When describing the Guarani people at the mouth of Orinoco, von Humboldt observed both the conditions that allowed them to remain independent (at the time) and the perils of colonial expansion:

> Even now, the Guarani owe the preservation of their physical and perhaps even their moral independence to the loose, semifluid bog soil across which they light-footedly move about, and to their residing in the trees: a high refuge to which religious enthusiasm will likely never lead and American Stylite. (von Humboldt, 2014 [1808]:36–37)

This passage concentrates a system view that includes a vast array of elements, from soil composition to human habitus, morals, and the role of religion in colonization. Alexander von Humboldt continues his observation of destruction brought by the presence of European ways of doing: "Since the discovery of the New Continent, Llanos became inhabitable by humans. [...] cities have been built here and there on the rivers of the steppes. All across the immeasurable space animal husbandry has begun" (ibid. 37). Unlike his contemporaries, he did not see this type of development as favorable to life in general but as gradual destruction with unforeseeable consequences. We can safely affirm that von Humboldt is the first thinker that, witnessing the rapid reshaping of the surroundings by human activity, understood the influence of human activity on climate and warned us about its possible dire consequences:

> When the charred grass falls to dust under the vertical rays of the sun, the hardened earth gapes open as if shaken by powerful tremors. [...] A hazy, almost straw-coloured half light is thrown by the seemingly low-hanging heavens upon the desolate plain. The horizon draws suddenly nearer. It constricts the steppe *and the mood of the wanderer as well.* The smothering heat of the air is increased by the hot, dusty earth floating in the atmosphere, which is veiled as if by fog. *Instead of cooling*, the east wind brings only more warmth as it blows among the long heated ground.
>
> The pools that the yellow-bleached fan palms protected from evaporation also gradually disappear. (ibid. 37–38)

Von Humboldt writes about the impact that intensive animal husbandry has on the environment at a time when environmentalism did not even exist. In the passage above, we can see, again, his attention to details and his capacity of seeing the world as an interrelated system, from the rays of the sun at the horizon to the feelings of a possible wanderer. Through his work, he also validated the scientific approach on nature and shaped our understanding of it and of climate manifestations. In her biography of Alexander von Humboldt, Andrea Wulf (2015) clearly shows the impact his work has on our contemporary approach on the environment:

> Ecologists, environmentalists and nature writers rely on Humboldt's vision, although most do so unknowingly. Rachel Carson's *Silent Spring* is based on Humboldt's concept of interconnectedness, and scientist James Lovelock's Gaia theory of the earth as a living organism bears remarkable similarities. When Humboldt described the earth as 'a natural whole animated and moved by inward forces', he predated Lovelock's ideas by more than 150 years. Humboldt called his book describing this new concept *Cosmos,* having initially considered (but then discarded) 'Gäa' as a title. (Wulf, 2015: 9)

Keen to draw historical parallels and preoccupied by the impact of human presence in general, the author identifies similar periods in European history, when the surroundings shifted their status from generators, providers, and shelters of life, human included, to rich-in-resources domains to be exploited, that is the establishing of agrarian societies. By comparing his observations on the South American continent with historical knowledge from other parts of the world, he was able to identify the effects of the modification of the environment that intensive agriculture is exerting. He explicitly linked this with a certain attitude toward the surrounding world:

> But one forgets that Southern Europe had a different appearance when agrarian Pelasgic of Carthaginian peoples first gathered there; one forgets that early human civilization pushed back the forests, and that the *drive of nations to re-create* has gradually robbed the Earth of the sylvan adornment [...] that is more indicative of the youthful period of our moral civilization than of any other part of history. (von Humboldt, 2014 [1808]: 158, my italics)

Here, "the drive to recreate" is at the core of environmental transformations. While von Humboldt equates "human civilization" with agrarian societies in the spirit of his epoch, he is nonetheless aware that this is not the only form of possible human organization nor the highest one. The drive to recreate one's own surroundings instead of blending into it is something that the author links to "our moral civilization," indicating once again his deep understanding of the entanglement of morals, ethics, and relationship with the environment. At the same time, he was highly critical of his society, the exploitation of the environment, and the political forms of organization that encouraged it—because they were based on it. Von Humboldt drew very clear parallels between slavery and exploitation of nature and opposed them vehemently. In *The Political Essay on New Spain*, he upheld his views on the necessity of independence from European rule that he saw as "barbaric." He linked his observations of geography with that of economy, politics of race, and religion, and, while calling upon the "culpable fanaticism" of the Christian missionaries, he admired the local people worshiping nature as "god." Andrea Wulf sums up von Humboldt's argument: "[...] colonial society was based on inequality; the indigenous people were neither barbaric, nor savages [...[and the future of South America was based on subsistence farming, and not on monoculture or mining" (Wulf, 2015: 180).

Today, in hindsight, we can see how premonitory his writings were but also that their cautionary advice was discarded in the past 200 years. The newly independent nations of South America, caught in the entanglement of global development, trade, and power dynamics, walked the path that von Humboldt warned against. Resource exploitation and intensive agriculture were—and continue to be—the exchange token that put a country on the global map. There must be something wrong with the map, if becoming part of it means destroying the very things that constitute one's particularities. This is why mapping as a method, understood in a more general sense of representing something within a system of signs with rather fixed significations, is worth a closer look.

Maps

Henricus Martellus was a map maker of German origin, living and working in Florence in the fifteenth century. In 1491, he produced a detailed map of the known world, today kept at Yale University, said to be consulted by Christophor Columbus for his voyage across the Atlantic. Columbus had hoped for the route to Japan and India, also basing his knowledge on previous maps and writings from the Arab world; the scientists in the period of the Abbasid Caliphate renaissance to the Mongol invasion (roughly from eighth to thirteenth century) had accepted the theory of the round Earth already and had based their trigonometry calculus on it and also their maps. The theory was not new; Greek philosophers already intuited the shape of our planet, and passed the knowledge on to India, from where it was subsequently spread into Central Asia and the Caliphate. Henricus Martellus himself used previously produced knowledge about the world, and it is particularly important that he based his description of Southern Africa on maps as old as those produced by Ptolemy in his Geography and on descriptions belonging to African native explorers as depicted by an Ethiopian delegation to the Council of Florence in 1441 (Cummings, 2015).

Martellus' map looks quite similar with the shape of the Eurasian and African continents as we accept them today. However, the descriptions of what makes up and who inhabits the lands is not necessarily accurate. For example, the "Palotii" people from Southeast Asia with ears so big as to use them as night covers are definitely beings of fantasy. But so are many other descriptions that recreated the territories at scale on the parchment and that shaped the imaginary picture of the epoch. It also shaped the perception of the world. At the time, a map contained everything, from relief to human geography, fauna, and various other types of information. They were undifferentiated. Gradually, maps became specialized and more exact about the specific detail they sought to emphasize, arriving to today's hyper specialization: topographic maps, meteorological maps, physical maps, political maps, economic maps, and so on. They became more and more exact about a more and more specific measurement. This means that all other elements that are contained in the represented area are also gradually discarded and with them their relationships. For example, we can have the elevation map of a territory, and we will need a different map in order to understand what makes up that elevation and yet a different one for the forms of life present on it. The specialized maps—or specialized representations—are very exact but in many of the cases almost unidimensional. The gain in unidimensional exactitude is achieved on the expense of the possibility of multilayered description and explanation. Jorge Louis Borges (1999) beautifully recounts this complicated relationship between exactitude, reality, and the accuracy of representation in his story, *On Exactitude and Science*. Only a paragraph long, the story shows how the ambition of creating an *accurate* map would require that the map is necessarily 1:1 scale and covers the entire territory it represents in order to represent *exactly* the materiality of its geography. Obviously, this renders it useless.

A map/a model therefore has to trade between *exactitude and accuracy* in two ways: by restricting the number of characteristics it depicts or by extending the scale all the way to a 1:1 ratio. The first suffers from a gradual loss of represented elements that will need to be inferred or ignored in the subsequent use of the map. So the map will be accurate but only for the elements it records therefore *not* exact.[1] The second is impractical, as Borges's story recounts. The story is a brilliant observation on the fact that what can be gained in exactitude is lost in power of description and relevance. In a more contemporary language, the measured elements constituting a map are called "data points." Maps are models based on data points with the limits of any model: it is a representation of the reality that is never exact but only an approximation of it—from the perspective of the purpose with which data sets are produced. Let us say that data are produced by exact measurement (a supposition that it is itself problematic), but no measurement is in fact exact. Moreover, once we start making inferences based on the measurements, they start acting as proxies. Sometimes they are good proxies. Sometimes they are not. On the relationship between proxies and the reality, they are standing in for, particularly on how they are used in the domain of artificial intelligence, Cathy O'Neil (2016) has the clearest critique on this matter in *Weapons of Math Destruction*. Her core argument is that the proxies used to create an image of the reality and that feed the algorithmic models are imbued with assumptions that are not recognized as such.

The accuracy of a measurement that creates a data point depends a lot on two factors: first, the suppositions embedded in the relationship between the measured element and the reality it represents. All measurements are standards of "ideal conditions," meaning that any factors that may influence the measurement in the moment of producing it are, willfully or not, ignored.

Second, there are the suppositions embodied by the measurement itself. If we analyze only the meter standard, we understand the level of both concreteness and abstraction that European scientific methods imply: a meter is a very concrete measure and a very particular one. Established by the French Academy of Science only 11 years before von Humboldt's voyage, the meter standard was developed in France as the one-tenth million fraction of the distance between equator and the North Pole measured on the meridian that passes through Paris. It is intimately linked with Paris and its particular geographical position. It sounds very concrete, but the meridian is an imaginary line, recreated through calculus and abstracting any other particularities of the land—in fact it assumes the uniformity of its topography. Like any model, it is an abstraction.

However, once the meter becomes standard, the entire world measured by its means automatically refers back to Paris, the location of which is anything but arbitrary, but which becomes the measure of universal validity of measurement. This is a clear departure from King Henry the First of England's foot that stands for that homonymous measure or from the "three grains of barley, dry and round, placed end

[1] Next chapter will offer a detailed history and critique of the assumption that lies at the foundation of data production and the belief in the objectivity of AI algorithmic models: the fact that exactitude is the same as accuracy.

to end lengthwise" that king Edward the Second decided to mean "an inch." This shift of reference was a step further into transforming our view of the planet from something to which our bodies are organically linked, to something akin to a self-referential lab, an object which we happen to inhabit but with which we are in a remote relation.

Fascinated by the scientific methods of measurement that were developing in Europe, von Humboldt intended to go and observe the so-called new world with these instruments.

In 1802, he was carrying with him on his ascent of Chimborazo volcanic peak a series of measuring instruments: "[...] he had left most of the baggage behind, but had packed a barometer, a thermometer, a sextant, an artificial horizon and a so-called cyanometer with which he could measure the 'blueness' of the sky" (Wulf, 2015:2). His keen mind and meticulous methods of observation allowed him to understand, for example, how altitude and latitude are somehow corresponding to each other in climate manifestations. That is, going up or going away from the Ecuator toward North or South would have similar effects on climate, flora, and fauna. Observing, measuring, and cataloging, he and his companions generated a vast array of knowledge regarding our environment, including the concept of environment itself. His very voyage influenced Darwin's decision to embark on the HMS Beagle and set out for the trip that changed the vision of how humans evolved, replacing us within the animal realm in the creationist-infused Europe of the time.

Later in the very century, the book was published, the desire to legitimize the knowledge about humans with scientific discourse pushed toward positivism and further to behaviorism, which both cut off humans from their environment and discredited for many decades to come the ideas surrounding the holistic approach on societies. These currents are at the very core of the separation not only of humans in general from the environment but the separation between the "modern' human—conceived as an individual that behaves rationally independent of the surroundings—and "the others," the "savages" at the mercy of nature's elements and caprices.

Measurement was and continues to be at the basis of scientific observation and methods. The relation between the particulars of the phenomenon observed and the transformation of the observations into a generally valid explanation that perfectly mirrors reality—that is the problem of induction as first framed by Bacon in the seventeenth century—remains itself problematic. The next chapter will be dedicated to how this problematic relationship traversed time and came to be embedded in the algorithmic view of the world that characterizes the AI-infused practices. Von Humboldt's century first proposed statistics in order to solve this conundrum, and the solution is perpetuated to these days. A sufficient number of observations on any phenomenon, assigning them mathematical values (numbers that is, as Leibniz first suggested), and calculating statistical values of these numbers such as average and median, would reveal the bare facts—the modern facts, as Mary Poovey (1998) calls them in her book on scientific epistemology, *A History of the Modern Fact*. This statistical solution in establishing truths about our world continues to be used, and it is not a bad model of approximation, particularly for some domains of knowledge, in some of the sciences. But it morphed into what today is called the quantitative method, and it dominates *all* discourses about reality. Its main problem is the claim to pure, "real" objectivity. This claim, if pushed to its logical conclusion that is embedded in it, disregards all the particular cases that do not fall into the statistical model generated by the observations.

Alexander von Humboldt's writings, including the *Political Essay of New Spain*, were packed with measurements, hard data, and numbers that he used to sustain his arguments. However, von Humboldt's arguments were not built on the quantitative method but on the combination of numbers and observations which today would fall under the qualitative methods. He wrote too early into the century to have his observations, and perhaps his inquiring mind, be hijacked by the fever of positivism that turned statistical probability into "scientific" certainty. I put the word scientific in quotation marks here because scientists know they deal with uncertainty and approximation. However, most of the public and a significant part of those using quantitative data for their models that justify actions invoke the presumed exactitude of "science." This presumption was forged in the past 200 years. Von Humboldt used the data measurements in order to make an argument against exploitation and even against the idea that the "wealth of nations" could be revealed by numbers. But the methods themselves became more important than any argument. *Moreover, the methods became the argument itself.* In other words, the numbers do not anymore express a variable and probabilistic reality, but they replace it. The expectation is that the reality itself "behaves" rationally, as numbers do.

The Lab

The scientific leaps in the past 200 years also meant refining the instruments of observation, bringing a variety of sciences to play into our understanding of nature, human nature included. Along with this, it has also brought the development of the system of knowledge that has at its base the model of the laboratory that tends by design to separate between the elements observed. This separation comes from two directions: the specialization of different sciences and the need to legitimize them as stand-alone disciplines on the one hand and the method itself of observing discrete phenomena or their components on the other. The mechanistic view of the world predominated the positivist era of the nineteenth century; its fundamental belief is that the whole can be explained by the functioning of its elements, hence the focus on constituting parts and their measurements.

But data needs to be created; things need to be measured. In order to be measured, they need to be brought into being. Sometimes the measurement itself brings them into being; sometimes they need to be insulated in the abstract space of a lab in order to be measured. Mostly, in order to measure anything, first we recreate it according to our interest and separate it from the rest of the system. This way of knowing, that seems natural throughout modernity, is applied today to both human and natural phenomena. It permeates our ways of knowing and ways of relating to the world. It is based though on a creation of a vision of the world constituted from discrete measurable parts that somehow stick together and form the world. It is thus possible to generate massive amounts of measurements, that is, data. It operates a fundamental break in the flow of things, so to speak. Why? Because the data analysts tend to interrogate data as discrete, yet revealing facts. In one interview with a product manager from a software company, she told me how the appeal to data is used as the ultimate argument, everything else, such as qualitative assessments being discarded as mere opinions. The shortcoming was obvious to her:

> A dating app data revealed to them that they need to attract more men into the fold, because men are the ones who tend to pay for subscriptions. So all the efforts were concentrated towards this goal. It took some time to realize that in fact the solution was to attract more women onto the platform, this being the incentive for men to join. Of course, this was not in the data. Big companies start slowly realizing that data is not enough, or it may not even have an answer. But in disputes, there is always the final argument: data tells us. Data reveals that. We have the data to sustain this claim. It's exact, it's mathematical, or so the argument goes.

This type of thinking extracts a fragment from a phenomenon, renders it measurable, and then represents it as being the truth about that phenomenon. One may then concentrate on discrete elements but more often than not lose the glue that keeps them together. For example, rain may evoke a certain mood, color, or sensation of change in the atmosphere and a certain change in birds' behavior in different places or locations. It may well evoke thoughts of crops, hope, and the promise of life. It may be announced by a dragon, too, an ex-snake with a pearl in its mouth. But when we think of rain in a measure-infused culture, we think of water falling from the sky and perhaps about the quantity of it as well. We may think of the probability of it occurring expressed in percentage chances of rain—as suggested by our forecast screen apps. We may think of its acidity. If we live a leisurely life, outside the constraints of working schedules, we may think of sleeping, but this is wishful thinking. What falls through the cracks, when rainfall is reduced to the measurable parameters that constitute it, and reconstitutes it as a knowable phenomenon within this very system? What externalities are we eliminating? Those externalities are the very relationships between the elements measured and those overlooked by measurements, too. It is their influence on the phenomenon itself that we fail to recognize, as we fail to recognize today the myriad links between every little part of our planet, despite the fact that von Humboldt and some of his intellectual inheritors emphasized them and continue to do so. Jakob von Uexküll, remarking the way we use technology not to replace us in the natural realm, but to further separate humanity from it, emphasizes this as an effect of the mathematical tools we use in knowledge production: "Much lauded human technology lost all sense of Nature; indeed, it boldly presumed to solve the deepest questions of life, such as the relationship of human beings with God-Nature, with its wholly inadequate mathematics" (von Uexküll, 2010 [1934]:191).

Today, we rely on massive amounts of data, and we imagine that having collected it is the same as knowing what happens in nature. But the most important thing we tend to forget is that data is not collected, it is *produced* at the intersection of the instrument with the measured phenomenon. Data is a *result of a complex interaction*, reduced to a simple number. Knowing the speed of wind and amount of precipitation in a storm leads us to create complex classifications that allow us to characterize the phenomenon and to assign numbers to its tail: a number 3 Hurricane or a number 4 typhoon—signs that bear the mirage of management and control. They almost become lullabies; we know what we can expect. This is a very particular type of knowledge that has been passed down to us from the tradition of modern Western thought and that it is based on the presumption of predictability and complex probabilistic causal chains. From the Enlightenment on, numbers, rationality, and nature have been inextricably linked in Western scientific thought.

Climate change and the anthropic influence on it, so clear to von Humboldt, has its own measurable parameters and set of data that can also be used as proxies. But climate change physicists who study the atmosphere and climate change are the first to tell us that we really do not know, or understand, the possible implications of our actions. Sometimes they are surprising. Sometimes the modifications of values that we use for understanding climate change are foreseeable within a range, but we do not know the effects of this on the proxies we used them for. For example, we know now that even if we stop all emissions of CO2 in the atmosphere, the warming will continue as a result of climate inertia. We also know that between the 1940s and the 1980s, the high levels of particle pollution known as aerosols released above Europe and the North American continent in fact reduced the speed of heating up of the atmosphere "hiding" the effects of climate change. The microparticles in the air reflected back solar radiation and facilitated the creation of clouds. The clean air act and the deindustrialization stopped this tendency, and so we are seeing rapid warming today. This was a very localized phenomenon, and it is reproduced at a different scale now in South and Southeast Asia due to industrialization and use of carbon combustion. We do not know the type of cascading effects that can be triggered at any time by the gradual warming of the oceans or by the possible stopping of the Atlantic deep current that works as a conveyor belt between the Arctic Circle and the equator. But we know that the oceans absorb *every second* the amount of heat equivalent to that of the detonation of five atomic bombs of the type used on Hiroshima, continuously 24 h every day.

How does this knowledge impact us? How does the framing of this knowledge in numerical terms influence the way we relate to it? I think herein lies a very important question, seldom asked. It is not the lack of knowledge, not even necessarily the way it is produced. It is the way we relate to it and to its production. It is our beliefs about the form and the forming of knowledge, together with the frame or reference of actions that we endorse that renders that knowledge both valuable and problematic, to say the least.

In the past decades, efforts have been made, some with remarkable results, in order to reduce, as much as possible, the sources of CO2, and to sequester CO2 from the atmosphere in the attempt to keep the world from heating up more than 2 °C by the end of the century. The efforts intensified and the technological progress allows for an industrial level of CO2 absorption from the atmosphere and its sequestering it in rocks—with a giant plant being built in Notrees, Texas by a company called "1pointfive," a reference to the temperature goal set for the end of the century.

Project Drawdown is a complex initiative that delineates possible solutions that, if applied in concert and on a planetary scale, will possibly reduce the heating tendency of the atmosphere, and make us meet the goals of keeping our planet cool, within the livable parameters for life as we know it. Many of those solutions that refer to multifactorial interventions, from transport and mobility to efficiency of energy production can be, and are, optimized by the use of specific AI systems. There is a caveat though.

Kate Marvel is a senior climate scientist in this project and a physicist by training. In an interview with David Wallace-Wells for the Ezra Klein Show (Podcast, n.d.), she explains quite clearly how human response to climate change can create effects that are even more unpredictable than the current scenarios without it. She also explains how the scenarios themselves, albeit based on an important amount of data, are filled with incertitude and why this is normal from a scientific perspective—since sciences did not forget about indeterminacies and interrelations, instability, and complex systems. With complex systems, there are always second- and third-degree consequences. In other words, even if we optimize to the max public transport, and energy consumption, we still do not know what *the effects of the effects* of these optimizations will be. However, the public's perceptions and its unrealistic expectation of both exactitude and accuracy from sciences (or better said, form "science") make that the approach on matters of climate change add to the complexity of the responses themselves. In *Politics of Nature*, Bruno Latour (2004) explains how in the current epistemology or what he calls the bicameral model of based on the platonic vision of "the idea" of a thing and its imperfect embodiment in the world, there is a confusion between sciences—the domains in which scientists specialize and work—and "science" that is the perfect idea that is never embodied but taken as etalon. The expectations of the public, and its understanding, is that sciences should behave like the ideal "science."

What happens when we relate as laypersons (and politicians are laypersons in regards to climate change physics) to nature as a measurable entity in a quasidigitized world, within a referential system that emphasizes measurement and control and it is permeated by the illusion of accuracy? The same thing that happens to our bodies when they disintegrate into sets of data that are then displayed on our portable screens. Both nature and our bodies misleadingly become manageable resources to be exploited in order to obtain other things (usually profit, pleasure, and power over others and oneself). What is left out is that, absent measuring, our bodies are ourselves, and we are nature. It is important to revisit the historical dynamic and the politics that link data with materiality in order to understand the attachment to the data driven society, and this is the focus of the next chapter.

References

Borges, J. L. (1999). *Collected fictions* (Chapter: On Exactitude and Science). Penguin Books.

Cummings, M. (2015, June 11). Hidden secrets of Yale's 1491 world map revealed via multi-spectral imaging [Article]. *Yale News*. https://news.yale.edu/2015/06/11/hidden-secrets-yale-s-1491-world-map-revealed-multispectral-imaging#_msocom_1 (Note: This is listed as an article because it appears on a news website, not in a scholarly publication).

Graeber, D., & Wengrow, D. (2021). *The dawn of everything: A new history of humanity* (1st ed.). Farrar, Straus and Giroux.

Hernandez, J. (2022). *Fresh banana leaves: Healing indigenous landscapes through indigenous science* (1st ed.). North Atlantic Books.

King, T. (2008). *The truth about stories: A native narrative*. University of Minnesota Press.

Latour, B. (2004). *Politics of nature: How to bring the sciences into democracy* (1st ed.). Harvard University Press.

Mueller, M. L. (2017). *Being salmon, being human: Encountering the wild in us and us in the wild* (1st ed.). Chelsea Green Publishing.

O'Neal, C. (2016). *Weapons of math destruction: How big data increases inequality and threatens democracy* (1st ed.). Crown – Crown Publishing Group – Penguin Random House LLC.

Poovey, M. (1998). *A history of the modern fact: Problems of knowledge in the sciences of wealth and society* (1st ed.). University of Chicago Press.

von Humboldt, A. (2012). *Political Essay on the Kingdom of New Spain: Containing researches relative to the geography of Mexico... the physical aspect of the country, the... and commercial industry* (1st ed.). Gale Ecco, Sabin Americana.

von Humboldt, A. (2014). *Views of nature* (1st ed.). The University of Chicago Press.

von Uexküll, J. (2010 [1934]). *A foray into the worlds of animals and humans. With a theory of meaning*. University of Minnesota Press.

Wulf, A. (2015). *Invention of nature: The adventures of Alexander Von Humboldt, the lost hero of science* (1st ed.).

Podcast: Ezra Klein Show (n.d.). (https://www.nytimes.com/2023/09/05/podcasts/transcript-david-wallace-wells-interviews-kate-marvel.html)

Chapter 5
Data, Facts, Reality

I remember a running joke from my childhood that made us invariably laugh: someone enters a train compartment in which three people were already riding. He greets everyone and takes a seat. One of the people previously present utters a number: 5. The other two burst into laughter. The episode repeats with each one taking turns in saying apparently random numbers that made the other two laugh. The newcomer, completely baffled, asked: "What is going on?" They answered: "Well, you see, we know the jokes by numbers. So in fact we are telling jokes to each other." The newcomer waits for a while and then utters: 17. The other three look at each other knowingly, but they do not laugh. So he asks: "Why aren't you laughing? Is 17 not a joke?" The answer comes: "Of course it is. But it matters how you tell it…" Leibniz's dream of a perfectly intelligible world obtained by assigning numbers to significations, a syntactic approach that is, is shattered by a children's joke that brings back semantics or the notion of context with all that it implies. But not everyone knows the joke or understands it for that matter. Well it depends how it is said after all. I will take a shot at it in this chapter.

The people in the joke transformed their stories into data and codified them. They operated on the assumption that data reflects reality, forgetting that both the production and the conditions of codification of data are context-dependent, as is their decoding. The punchline brilliantly reveals the multiple layers of meaning and potential misunderstandings that escape if the process relies massively on data. It also reveals that those familiar with the production of the data know, consciously or not, that data is not everything.

But data is also known as the oil of artificial intelligence, and the unquestioned belief that data is objective is the basis of the indiscriminate adoption of AI. This belief reflects in the language surrounding data. Oftentimes even people engaged in research talk about "gathering data"—as if data is preexistent, laying around ready to be reaped (just like any harvest or like oil for that matter). The truth is, data is generated by the interactions of the tools with the environment and produced to be

A. Balasescu, *Climate Change in the Age of Artificial Intelligence*, Artificial Intelligence, Simulation and Society,
https://doi.org/10.1007/978-3-031-90042-6_5

intelligible by machine systems in processes of "cleaning," quite similar to oil refinement. However, data is infinitely produced; there is no scarcity of it, and this generates a very different general economy than oil. The extraction of oil is dependent on the presence, on the preexistence of oil in a certain place. The extraction of data is only preceded by the presence of the tools that render the interaction into data. The logic of extraction is thus perfectly independent of the preconditions of the environment. In her book *The Mushroom at the End of the World*, Anne Tsing (2010) explains how sugarcane plantations served as the model of colonialism through their possibility of scaling. Given a very simple set of conditions such as proper climate and availability of workforce (made possible through slavery), any patch of land could be turned into a sugarcane plantation. The plant will extract the richness of the soil, the light of the sun, and the soothing cycles of the moon and transform them into a sweet fiber from which the forced labor of the slaves extract the commodity to be sold for exorbitant prices. The costs are unacknowledged; the depletion of the soil and the destruction of biodiversity to leave the plantation are not parts of the accountants book. The inhuman slave labor is infinitesimally cheap when compared to the profits. And the scalability is practically limitless, given the proper climate conditions. Data extractionism brings the scalability one step further by eliminating both the climate conditioning and by reducing even more the data production costs. The conversations regarding the fact that people are data producers by simply using the apps and that they should be compensated for this are well known. They also lead nowhere, really. This is because they are taking place within the same regime of economic principles, and the argument against that compensation is that the very use of "free apps" is the compensation. Looks like a fair exchange—produce data and use the app. There are two matters left unacknowledged here: the nature of that data, obscured by thousands of pages of agreements notices that nobody is really reading and everyone signs, and derived from this, the enormous scalability of its value within the information economy. The data generated by using one app is enormously more valuable within this system than the possibility of using the app free of charge. However, we did get used to the freebies already.

And another "however," the value of the data a person generates (or any data) is based on both the fundamental belief that *data perfectly reflects reality* and on the capacity to transform it into "information"—that is to refine it (if you are to keep with the oil metaphor). The caveat is that free access to data is meaningless, absent the refining power, as free access to brute would be, too. As Andrew Zolli from planet.com observed in one of our conversations, the promise of free access to data is not only empty, is misleading and probably dangerous absent regulations about who else have access to it, beside the person who generates it: "If data is information and information is power, we need to think who has the capacity to transform data into information? So, if data would be made free for everyone, the giants with this power would be further immensely powered, while all other people would only be marginally empowered, and the gap would continue to increase." I am afraid this already happens.

This chapter invites the reader to discover when in history numbers came to mean "truth" and how data, facts, and reality have been brought together under the roof of natural laws, with the simple assertion "this is how nature works." The quasireligious undertones of this process is captured in Bartoletti's (2020) syntagm: "the sanctity of data." Up to now, the history and the core of this belief have rarely been questioned, despite numerous calls to both address bias in algorithm design and for more ethical use of AI.

Reality Is Not What It Used to Be

The ethical dilemmas surrounding AI have seldom been connected[1] with this unshakeable belief that numbers are perfect mirrors of reality and, by implication, of natural phenomena. However, the belief in the objectivity of data that implies an objectual world manageable at all levels is at the core of this issue. This same belief is also at the core of Silicon Valley culture and lifestyle, making it desirable for the rest of the world. It perpetuates itself inside AI labs and outside in the gyms, coffee shops, health stores, and hangouts that reflects to the letter the concept of manageable bodies performing in a managed "natural" environment. There are few ethnographic, autoethnographic, and memoir-like accounts of this particular culture; the most well-known recently is Anna Wiener's memoir *The Uncanny Valley* (2020); however the preoccupations with this particular area are much older, as Ellen Ullman's book first published in 1997, *Close to the Machine* (1997), attests.

Through historicizing this belief, this chapter brings the previous conversations to a conclusion, while constituting a segue into the second part of the book, which will analyze how data facilitates decision-making, what are the consequences of decision-making based on data, and how the drive for optimization has permeated most aspects of life, transformed its meaning, and changed our politics.

It is not technology turning evil; it is that our sociocultural current structures and norms are geared toward practices and values that also have an "evil" side. If I can quote Shakespeare from memory, "There is nothing either good or bad, but thinking makes it so" (Hamlet, act 2, scene 2); but technology does not have—for the moment, and perhaps never—the "thinking" needed to distinguish between a face of a coin and its reverse. Its constant and perpetual application only brings them both to their logical conclusions. Utopia and dystopia are living simultaneously, particularly if we think that this particular technology is made to transform everything into a one-sided surface that constantly produces data: a gridded pattern with

[1] There are of course notable exceptions quoted extensively in this book, such as Meredith Broussard, Kathy O'Neil, or Meghan O'Gieblyn. For the moment, the AI economy continues to operate with this assumption. Of course, one needs to always keep in mind the benefits of AI in sciences, from medical research to climate-related science, as well as its constant and rapid advances. The human protein folding patterns have been recently solved by "AlphaFold" an AI system whose designers received the Nobel Prize in 2024.

nothing underneath. Better said, everything that is underneath the data-generating grid is "left behind" in irrelevance. The byproducts of data extraction, the human lives themselves, that is, are residues of these extraction processes, just like the leftovers of sugarcane after the sweet juice is crystallized in the desired product.

How many times have you heard this? "If you do not learn the new technology you will become irrelevant, you'll be left behind" (from salvation?) and so on. Perhaps in fact this means, inside the internal logic of this discourse and keeping with the metaphor, that you will be left underneath the grid, thence also the movements that embrace "living off the grid." The question remains as to what are the costs, human, social, and nonhuman, of being left, leaving, and living "underneath"—by choice or by force.

Adrian Preda is a contemporary artist from Romania who approached this theme in a stunning visual manner. He produced a series of paintings depicting large marine mammals pushing through the surface of a net(work), as if trying to emerge at the surface of an ocean. However, the surface is inorganic and highly elastic, bending up and creating an inescapable environment. Each painting is titled similarly: the name of the mammal "trapped in space-time" (Fig. 5.1a, b).

If the whale cannot break through, we can imagine that it cannot breathe, either. The animal, the organic mass that strives to pass through but is caught (inescapably?) in the fluorescent grid pattern of the network may represent life itself, caught in the process of data production and extraction. Or, perhaps better said, we witness the confrontation of a certain idea of nature—that which is alive—with the inorganic world of technology, figuratively shown as the grid of space-time, that is, the

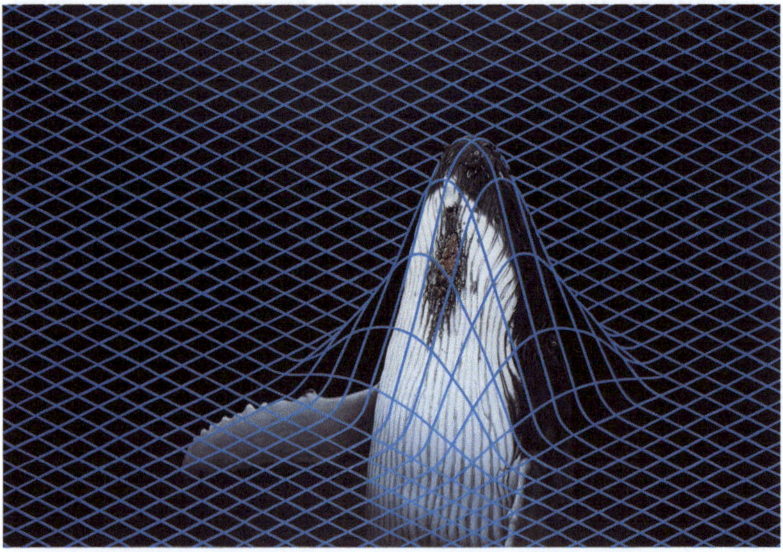

Fig. 5.1 (**a**) The whale trapped in space-time by Andrei Preda. H'Art Gallery by Dan Popescu. (**b**) The Narwal trapped in space time by Andrei Preda. Courtesy of H'Art Gallery by Dan Popescu

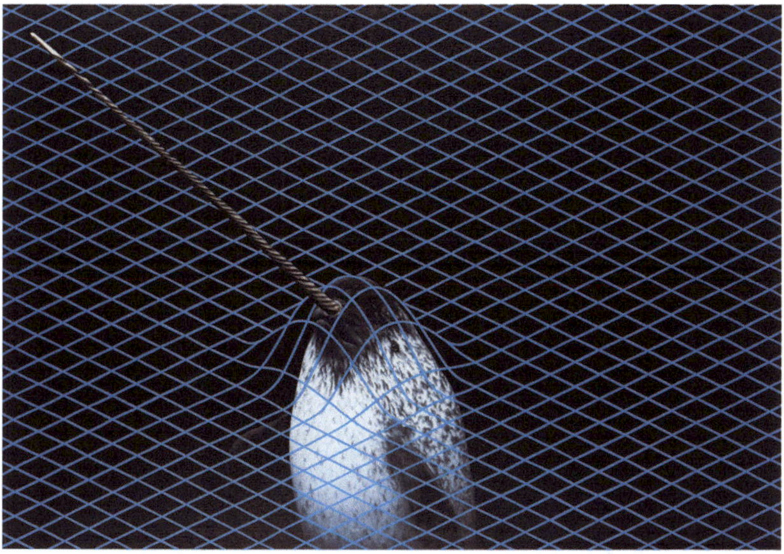

Fig. 5.1 (continued)

technological translation of the universe into mathematical language. If the whale in the painting represents nature, it could only pass through the grid if fragmented into the squares of the grid that makes it translatable into data.

Many climate change tools based on AI, with all their undeniable usefulness, propose just that: the creation of ever so finer grids of data extraction in order to recreate and represent the global climate system in an ever more granular manner. Interestingly enough, visualization is a big part of this intelligibility. In the AI for good series of conferences online, I assisted at numerous presentations that focused on precisely how to create a higher resolution imagery of weather patterns and movements. As if "seeing is believing." As if better seeing is also explaining, and if explained, then perhaps it is also possible to manage or at least to identify leverage points of action to manage the outcomes.

So this subtracting and abstracting in the realm of data is allegedly done with the purpose of making the environment more intelligible, and thus, it is implied, more predictable. There are two questions emerging straight from these declared purposes: intelligible by whom, or by what kind of intelligence, and predictable in what sense, since there is a general consensus that the global climate system is characterized by high unpredictability?

A few important things happened in the post-Renaissance period that shaped the way we relate to our world and its intelligibility. One of them is the complex set of transformations brought by the introduction, in the fifteenth century, of the so-called Arabic numerals, in fact originating in Hinduism, with their star number "0." This sign is meant to mean "nothing" but also it is the origin of everything as it generates a system that accepts infinity as possibility.

The changes potentiated by the usage of this sign impacted the most unexpected aspects of life, from banking and the idea of money and value, to spatial representations, architecture and urbanism, and ultimately the relationship with the idea of "self," in his book, *Signifying Nothing*. In *The Semiotics of Zero*, Brian Rotman (1993 [1987]) explores and synthesizes these entangled changes that impacted both thinking and practices, from the Renaissance on, leading to what we call the modern period.

An arithmetical sign, zero made possible a few transformations in the way we relate to the material world. Rotman argues that zero is a *meta-sign* that has an ambiguous role: it is both the originator and a participant in the system. But prior to zero, numbers referred directly to things, they were associated with a finite number of things generating a finite number of signs. This is *arithmoi*, the concept of counting present in ancient Greek philosophy that survived until zero appeared, and with it *algebra,* an Arabic name designating calculus within the system in which signs and signifiers are forever separated.

> To make zero the origin of number (sic!) is to claim for all numbers, including the unit, the status of free, unreferenced sign. Not signs of *something*, not *arithmoi,* certainly not real collections, and not abstraction of 'units' considered somehow as external and prior to numbers, but signs produced by and within the arithmetical notation. (Rotman, 1993 [1987]: 29)

The introduction of zero influenced not only arithmetical signs but also "picture signs" (drawings, paintings, representations) and of course money signs. It did this by producing a system in which an infinite number of signs can be produced—therefore an infinite number of numbers, pictures, or transactions. While this system is thought to represent (name, draw, price) an a priori reality to a preexistent knowing subject ("the one who counts," "the one who sees," "the one who buys"), Rotman argues that in fact accepting and operating in this system reverses the relationship between signs and things they are supposed to signify.

> The system allows the subject to enact a *thought experiment* (calculating, viewing, dealing) about this reality through the agency of a *meta-sign* (zero, the vanishing point, imaginary money) which initiates the system and affects *a change of codes* (gestural/graphic, iconic/perspectival, product/commodity). [...] What are taken to be pre-existing numbers are given names, scenes from some supposedly pre-existing visible world are depicted, goods conceived as existing independently of, and prior to, the agency of money are assigned a price. [...] And it is this double ambiguous role played by zero, by the vanishing point, by imaginary money, which ultimately destabilizes the scheme presented here and deconstructs the anteriority to signs this reality is supposed to enjoy. (Rotman, ibid: 27)

The infinite as concept moves from the potency of divinity whose infinite power created a finite world, to the possibility of creation of various infinities within the world system. Thus, the infinite power of divine creation is brought into the realm of humans if they operate in a system that allows the power of an infinite creation of signs. The zero, the vanishing point, and the imaginary money are the privileged position of the human creator, placed not outside the system, but within it. But with this, the relevance of the a priori existing reality is diminished, since the sign does not need anymore the materiality of the world as ultimate referential, the world of numbers being a self-generating world. The material implications of postulating the primacy of immateriality, an illusionist act after all, are vast and defining for the way in which we organize our sociopolitical and cultural landscapes to this day.

First, the creation of the rule of perspective in drawings is obtained by the introduction of the vanishing point in bidimensional paintings and sketches. All of a sudden, drawings acquired a three-dimensional quality, offering the illusion of depth. Previous to this method of drawing, depth in bidimensional representations was suggested by drawing the objects smaller or bigger in the frame. This is a rule that we still find today in Byzantine religious drawings or in Persian Miniatures.

The Italian Renaissance painter and, not incidentally, mathematician Paolo Ucello explicitly depicted this shift in representation in his painting *St. George Slaying the Dragon* (Fig. 5.2). The scene shows a riding knight, coming into the frame from the right side, charging a dragon in front of a tent. On the left of the frame, at the back of the dragon, there is a feminine character, let us say a princess, and in her background a castle. The background of the right half of the painting that contains the knight is drawn using a vanishing point, while the half in which the dragon and the princess appear follows the old rule of perspective. The effect is stunning, and contemplating this painting, we can meditate on how much our perception of both depth in paintings, and also of the self, changed with the generalization of this rule of representation.

At a careful examination, the painting may also lend itself to a layered interpretation: the Dragon is standing between the knight and the princess. The knight is completely armored, faceless, and almost a mechanical being on a horse. On the one hand, we may interpret it as it is customary: the knight is saving the princess from the Dragon's hold. However, the princess does not look in distress. On the contrary,

Fig. 5.2 St. George Slaying the Dragon, by Paolo Ucello, courtesy of Musée Jacquemart-André, Paris—Institut de France © Studio Sébert Photographes

hers and the dragon's side of the painting is bathed in a soft light, while the right side is darker. What if the Dragon is in fact her protector? Or if they protect each other for that matter? Knighthood precepts are pretty much bound to female domination: honor, righteousness, chastity, (over) protection of territory, etc. This specific painting may as well mean this: the new order, the nascent era of mathematical Enlightenment leading to the future modernity and technologically reinforced patriarchy is advancing to take possession and control of the female body. The Dragon, the exponent of the mysteries and fantastical creatures of the enchanted world that is leaving the history's stage, is about to be slain.

Second, there is the invention of imaginary money, at first meaning paper money, but continuing to our current cryptocurrencies. This meant the divorcing of value from the metal and its replacement with the promise of redeeming the value inscribed on a paper by the guarantor of that paper: at first banking institutions. Today it is mostly national banks (at the time of the first imaginary money nation-states were still nonexistent), with few exceptions. For example, the Hong Kong dollars are issued by three commercial banks: the Hongkong and Shanghai Banking Corporation Limited (aka HSBC), the Bank of China (Hong Kong), and the Standard Chartered Bank (Hong Kong) Limited. They operate under the authorization of the Hong Kong Monetary Agency.

The cipher "0" made possible the invention of the balance sheet in banking and the introduction of the representation of credit as a negative balance. It also made possible the invention of the merchant recording based on the double entry booking practice with enormous cultural consequences, impacting the very fabric of our perception of materiality (and reality) through the invention of the "modern fact"—as Mary Poovey (1998) calls it.

Drawing as Worldbuilding

First things first. The vanishing point in drawings is obtained by establishing one (or sometimes two) points of reference and tracing radiant gridded lines that converge to that point. The objects in the drawing are then aligned and proportioned to follow the grid thus obtained. This led to the idea of the centrality of the viewer, by placing the painter, and implicitly the viewer, at the generating center of the depth of the image. The viewer becomes thus part of the drawing and also has the illusion of the generating point of perspective.

This new practice of drawing had an impact well beyond paintings. Italian theaters started to use this rule in order to create imaginary backgrounds representing streets that offered a depth of view to the spectators. Sebastian Serlio who worked and lived in the sixteenth century is credited with this innovation in theater scenography, and his work further inspired the design and construction of real streets in the eighteenth and nineteenth century Europe, as Vidler (1978) argued. The invention of perspective thus implied major changes in urban design:

> For the laws of perspective were not only those of illusion, of depicting three dimensions in two, but fundamentally the constructive laws of space itself. Thus the street, subject to perspective representation in the ideal theater, was transformed by this technique and shaped by it. (Vidler, 1978: 30)

The impact of this transformation is understandable once we do away with the Kantian idea of space as a preexistent category we passively populate. Lived space for humans is a direct result of design principles, and our spatial perceptions are influenced by the reorganization of space following the rules of design—no matter which of those rules we follow. Lived space is a creation, not an a priori.

The operations of design and the construction of streets following the rules of perspective created the possibility of a vista point, and moreover, it made any point along the street a vista point. The rule of perspective afforded a privileged role to visual perception and at the same time decentered the place of the perceiver in the case of transposition of a two-dimensional plan in three-dimensional construction. For instance, while for the representation of a street on paper, the vista point is the place of the designer or of the plan viewer, once on a street built on such a plan, every place on the street offers the same overarching perspective, combining the quality of the panoramic observer with the position of equal participation in the landscape-spectacle of the street. Thus, the use of vanishing point renders every observer position (while admiring a drawing or walking on a modern street) a position of vantage point with equal potential of vision.

> The vanishing point is the anchor of a system which *incarnates* the viewer, renders him tangible and corporeal, a measurable, and above all a visible object in a world of absolute visibility. (Bryson in Rotman, 1993: 14)

The vanishing point privileges every vantage point, giving it equal importance while with the same move dethrones the privileged godly overarching perspective. The omniscient, omnivisionary monotheistic God whose unique perspective previously acted as unique referential is gradually replaced by making every perspective a point of origin of vision. The viewer is the knowing subject, whose position is all encompassing and at the same time it generates the field of vision. First as a feeling—as the viewer started to feel, literally, that she is at the center of the space, the generator of the visual field—then, and in parallel, as a practice of thought. The birth of the individual subject, the "modern human" or the meta-subject as Rotman calls it, whose perspective replaced that of an overarching monotheistic God as source of truth and accurate vision of the world is intimately linked to these convergent phenomena surrounding the introduction of 0. The process is continuous and it continuously reorganizes, to this day, our field of relationships, in other words our subjectivation. The subject is not what it used to be nor is reality.

An important step toward the current place of the *knowing subject* and its relationship with the known or knowable (potentially seen) reality occurred some 200 years after the Italian Renaissance. Following Bryson's (in Rotman, 1993) analysis of the paintings of Vermeer's *The Artist in its Studio* and Velasquez's *Las Meninas,* Rotman argues that they illustrate the moment in which knowledge produced by the subject started to precede reality; in other words the sign preceded the materiality of the representation. Both Vermeer and Velasques present to the viewer the place of the painter himself, who appears as if really viewed by the person who paints. In other words, the painter created an imaginary view of themselves painting. As opposed to Van Eyck's painting *The Arnolfini Wedding*, in which the painter is represented in the mirror placed at the position of the vanishing point, Vermeer moves the vanishing point from the "reality" depicted by the painted scene into the imagination of the painter. If Van Eyck uses the mirror to point out to the painter and to show him as the source of the representation, Vermeer puts the painter (himself) in the painting creating the illusion of a reality generated by his own imagination. So does Velasquez, and he is even more explicit in showing this as an illusionist act: in *Las Meninas*, the painter appears behind a curtain, mostly hidden from the painted views, but not from us. The viewers can see that this is an illusion, while they can also perceive themselves not anymore as full participants and examiners of the reality represented in the picture but as the generators of the illusion that precedes and replaces the depicted scene.

> If, as seems reasonable, the seated figure can be taken to depict Vermeer, then, because Vermeer could not have seen himself in this way, his painting represents a scene that he imagined. The scene was nowhere but in his head. But if a scene can be literally nowhere, have no 'real' location, what then is the difference between presenting and representing? The priority of things to signs which sustains the distinction seems to disappear." (Rotman, ibid. 35)

Thus, the vanishing point moves out of the painting into the painter/viewer's head, transforming them into metasubjects. Rotman calls this the closure of the vanishing point into a punctum, marking the reversion of the relationship between the sign and materiality: materiality does not precede the sign, but the sign is divorced from and creates (the illusion of) materiality. The process is still morphing today, and the current posttruth era can be both traced back to the early modern privilege of each individual perspective, followed by the emergence of (the illusion of) the sign that precedes the thingness and that of the metasubject. The referent is not only visual but that of any sign that participates in presenting and representing reality. The godly privileged vanishing point that became generalized is the privileged *source "0" of the truth*, of the narrative story that organizes reality. If the story precedes and creates reality, and any storyteller is a metasubject, then multiple and competing metarealities emerge.

One step further took place when the many sources of the multiple representations preceding the reality became (apparently) equally legitime. It took 400 years and the invention of social media that democratized the storytelling legitimacy to accomplish the promise of decentering the vanishing points: made any vanishing point a punctum, a generating source of truth, while divorcing truth from any external referent save its own source. Truth became metasubjective and self-referential, just like any individual walking on a street built on the rules of perspective.

As a Matter of Fact...

Let us return for the moment to imaginary money and double-entry bookkeeping and follow through how thinking about the material world has changed with the introduction of the self-referential system based on zero. The subtitle phrase is an interesting one in the English language: as a matter of fact. At a first glance, it seems to encapsulate the idea that facts are material. But, if we look deeper, it suggests that the matter is of the fact—and not of the world. The fact has a materiality, yes, but that materiality belongs to the fact and not to a preexistent reality. What can this mean? Is it just a language artifice, or is there something more about it?

If we are to consider that on the one hand we tend to privilege quantitative perspectives and demonstrations of a phenomenon, and on the other hand, as we have seen, the self-referential numerical system generates its own internally coherent and infinite number of possibilities, facts expressed in numbers may be separated from the materiality they (pre-)tend to signify, while creating a new materiality of their own. At least this is how their history started, Mary Poovey assures us in her volume *A History of the Modern Fact: Problems of Knowledge in the Sciences of Wealth and Society* (1998). The author in a way picks up where Brian Rotman left off and deepens the analysis of the successive historical periods starting with the introduction of zero and the revolution in Western mathematics as applied to commerce. Her core argument is that the relatively newly adopted number system started to be used as rhetoric technique[2] that legitimized both the probity of the ascending merchant class and the shifting from sovereignty as expression of moral power to wealth as moral legitimation and measure of sovereign power. The author looks very specifically into the practices of commerce starting with the fifteenth century, since this is the domain in which the new zero-based mathematical system and the imaginary money are used extensively. Poovey records the changes in practices that shifted the Western thinking in matters of politics, knowledge, and the use of numbers as a form of legitimacy. To do this, she starts with the analysis of the manual for creating a double entry ledger, written in 1494 by Luca Pacioli, *De Computis et Scripturis* (1878 [1494]) This contains a step-by-step instruction set detailing how to organize the entry and the exits of goods and money from an economic unit, be it a household or a company. The principle is simple: in a table with eight columns, the fourth column contains the narrative description of the good or asset. In the three columns at the left of the description the accountant marks the date of the transaction, and in the right four columns the writer marks, in numbers, the value assigned to the described goods. The open book would have on the left-hand page the credits and on the right-hand page the debits. Two major operations are implied in keeping a double-entry ledger.

[2] Rhetoric here is understood in its classical sense of a system of knowledge which displayed mainly through means of speech is meant to negotiate, maintain, and possibly enhance property and social status.

First, there is the assignment of value expressed in numbers to the narrative description of goods. This is the moment in which numbers gain a rhetorical quality—they literally speak. They become signifiers of the material goods and assets and start acting (that is speaking) in the accounting system; that is a concrete embodiment of the zero-based mathematical system previously discussed. While the reference to the materiality exists, the value assignment is somehow similar with the story at the beginning of this chapter: each good has an assigned value in the same way that each joke has an assigned number. The link with the materiality of the world is still there; however, the complex semantic in the description of the goods or assets is reduced to a number representing value (imaginary or not, this is open to discussion). Through this operation, the material world is represented and resignified in the precise realm of numbers—the closed system with infinite possibilities. But the precision of the numeric system implied, or gave the impression of, the accuracy of the description: "the double entry system seemed to guarantee that the details it recorded were accurate reflections of the goods that exchanged hands *because* the [mathematical] system was formally precise'" (Poovey, 1998: 30). In this precise and self-sufficient system though, syntax is what counts: the complex context of qualitative descriptions from the central column disappears in a number, and it has to be so in order for the system to function and maintain its preciseness. Precision and presumed accuracy, however, illusory, are the characteristics afforded by this system. In current terms, we would call this data codification.

The second crucial operation, made possible by the quantification of the goods into numbers, is the balancing of the accounts. That is, if the entries and exits are not equal, the account is "unbalanced." In order to balance it, a fictitious number is added to the column of credit or debit in order to show an equal number on the bottom of the facing pages at the end of the ledger. That number corresponds to no external description of goods/assets; instead it corresponds to the internal "need" of the system to be balanced. But why did the ledger need to be balanced? The balance suggested both the scale of justice and the symmetry of God's world, Poovey argues. Visually representing a balanced image in the double-entry ledger, on the two facing pages at the end of the book, indicated that "the image of the scale of justice was subordinated to the figure of God's order, as if the scales have been brought to balance instead of indicating a weightier argument" (Ibid. 55).

The supposedly divine characters of abstract numbers are already presented in early Christian writings of St. Augustin, who lived in the fourth and fifth century AC. In his Confessions (2024), the early saint reflects on the difference between sensitive numbers and ideal numbers, relegating the latter to the divine principles that govern the world. The numbers that express the things that we count, he argues, are separated from the abstract and universal laws and numbers that have no sensorial dimension: "we perceived with all the senses of the body the names that we use when we count things; but the principles of the numbers are altogether a different thing, they are not images of the counted things, and thus they are existing in absolute [divine dimension]" (St. Augustin, 2024: 463, my translation). It took almost a thousand years for the Western Christians to render the divine perfection into visuals. The introduction of 0 made possible within this framework of thinking the visual representation of divinity through mathematical formula: the numbers that exist *in absolute* and that express the "divine," later "natural," laws governing the universe. While the Islamic Renaissance in the nineth to twelfth century used the system of "ideal numbers," and thus inventing algebra, they associated it, together with the trigonometry, to poetry and the poetic representations of the world. For the thinkers of that era and region, divinity was not representable or decipherable for that matter. It took the filtering of Platonician ideas through Christian monotheism, and the adoption of "Arabic numbers" in Europe's learning and commercial centers, to make the business of God a mathematical affair.

The adoption of the double-entry ledger had two major implications: social, leading to the elevation of trust in the merchant class, and epistemological, accounting for how the rule bound system of mathematics seems to guarantee the accuracy of recorded material. Regardless if the deficit of balance is compensated by a fictitious number, with no referent to the materiality, in fact this is congruent with Rotman's thesis that within the newly adopted system, materiality is secondary to the sign, giving primacy to the numerical, quantitative expression rather than the qualitative assessment. "In double entry then, the precision of arithmetics replaced the eloquence of speech as the instrument that produced both truth and virtue" (idem).

It took more than 150 years from the publication of Pacioli's manual, major epistemological transformations, and a devastating 30 years war, for Leibniz to propose a double-ledger system for the entirety of existence. His proposition of assigning numbers to meanings in order to afford clarity to messages and avoid conflict and misunderstandings, based on the mechanistic view of the universe, did not come out of nothing. What fascinated and eased knowledge production was the brevity and preciseness of numbers, leading to the adoption of numerical language in matters of life, state, and society. Poovey details in her book the journey of numbers entering all realms of explanations, from state power that was now possible to display by calculating wealth, to scientific endeavors and knowledge about societies and humans: the English Royal Society adopted and adapted the rhetoric of numbers and practically created the concept of "fact." Emerging from the perception that the brevity and preciseness of numbers that "privileged quantification over qualitative description, the priority accorded to numbers tended to make details that could be quantified more pertinent than details that could not" (ibid. 54). The shaping of knowledge production in the centuries to come was haunted by the paradox intrinsic in the ideas about what constitutes a fact: both a separate collection of observations prior to, and separated from any theory, as Francis Bacon postulates and something that "cannot exist—in the sense of being meaningful—unless they speak to some relation, which is always implicitly theoretical" (ibid. 96). As Poovey argues, the solution to the perceived problem of induction was "solved" on the one hand in the scholastic disputes of the seventeenth and eighteenth centuries by simply "disavowing political economy's relation both to political arithmetic and to accounting" (ibid. 31). On the other hand, accounting became an instrument that morphed into statistics and used in administration starting at the beginning of nineteenth century, leading "modern students of accounting to assimilate even early modern accounting to what Michel Foucault called governmentality" (idem).

For observations pertaining to humans and societies, Hobbes' *Leviathan* (1982) is crucial in adopting the logic of the double-entry ledger and simultaneously presenting an argument that has nothing to do with it. He postulated that his observations on "human nature" have universal validity not because they were based on an inductive process of analysis but because he observed them from a position of universality—meaning that each human, including him, occupies the same position, having similar and equal interests. Hobbes attempted thus to solve the question of induction—how do we get from particular observations to generalization. His answer was by universalizing and equalizing the vantage point of the observer. This led to his understanding of society as a tridimensional drawing based on vanishing point rules that rendered individual positions equal, equal to the author's point of perspective, and thence universally valid. This also allowed Hobbes to postulate that, all units being equal, mathematical rules can be applied in understanding and evaluating "human nature," itself bound to "laws of nature" or "dictates of nature."

As Poovey observes, quantification functions "as an interpretive instrument: by translating incommensurable differences or qualities into something that can be quantified" (ibid. 108). In doing so with human individuals, Hobbes "used the methodological precision perfected in double-entry bookkeeping to create the effect of certainty" (ibid. 109). In other words, the precision of the instrument that afforded the merchant class to claim *accuracy*, when applied to knowledge about humans generated the conviction of *certainty*. That is, the illusion that human behavior is predictable with *certainty*, not because it is predictable but because the *method* used to evaluate it is very *precise*. What gets overlooked in this equivalence operated by Hobbes between method and phenomenon is what Hobbes himself overlooked: the observed particulars—some of which may have even contradicted the model, in other words, human particulars, their subjectivity/subjectivation, and their diversity of environments and motivations. This is because the human society he presents in *Leviathan* is not necessarily what surrounds him—which was at the time, as it is now, a society ridden with conflict and violence—but what *ought* to be. He was writing around the same time as Leibniz, and his thirst for social order, for predictability and consensus was quenched by the precision of the mathematical system. This operation of overlooking anything difficult to quantify constitutes the basis of what Poovey calls "the modern fact": a particularity that becomes evidence not because of an a priori quality but because it can be quantified and be interpreted *as evidence* in the light of a particular theory and within the precise mathematical system. One can say that slowly materiality receded in order to give room to facts. As facts are numbers, they are innumerable within the zero-based system; they can be infinitely generated, their validity being assured by the precision of the system. They can be conjured up out of nothing, like the balancing numbers in the double-entry ledger. As facts proliferated, materiality (and reality) receded, and the matter became of the fact more so than of, or generated by, the external world.

... Data

The question of induction however returns in different forms and remains somehow unsolved to this day. The formalization of current methods into quantitative and qualitative methods is one attempt to answer it. The pitfall is the tendency to favor quantitative data and the attempt to quantify any qualitative assessment, that is, to digitize.

When interviewing a product director from an international firm about data and actionables, she recounted to me the following: more often than not, in the process of making a product, despite the fact that the idea is there, the prototype may be even created, the team would not move forward if they do not have "data." Everyone is waiting for "data," and the moment the data arrives there is a frenzy and excitement around the fact of "having the data." In the interview, she recounts a moment in which two teams were waiting for the data to come. At some point, one of the teams receives the data sets and enthusiastically announces: "we have data, a lot of data, it's amazing, we got it!" Someone asked "what is the data about?" The answer came promptly: "we have no idea, but it doesn't matter. We have data!" The qualitative assessment is not considered valid, unless it is sustained by quantitative data. The reverse is not true. That is, if one presents a series of numbers and data about a subject, the numbers "speak for themselves" and are taken as precedent and fully explanatory of the reality that they are supposed to represent. I have already discussed one example in the previous chapter. Another telling one is that discussed by Mueller in *Being Salmon, Being Human*. The author recounts a conference about salmon farming and the arguments that depicted the ideal conditions that fish enjoy, expressed in numbers: cubic meters of water per fish available in a farm. At a first glance, and for the audience, the number was amazingly generous and thus convincing, probably even for the presenter. The skeptical author made a little bit more complex calculation introducing body mass and took into account how physical volume and depth play into the picture and transformed it into metric cubes of habitat for humans. The results were dire. It turned out that the numbers convincingly presented as a triumph for the salmon farming industry in fact indicated an equivalent of humans piling up on each other in very small spaces. His entire book deconstructs, using this example, the illusion of the modern fact, that is the illusion that *numbers have an immediate, accurate, and certain* relationship with the phenomena they are meant to describe. In fact, they only are *precise* within the closed system of mathematics. But, as we say in anthropology, it is not as much relevant if a belief is "true" as it is the question of what are the effects of the belief on the societies that hold it.

During my PhD years, I had a friend from the behavioral studies department who was working on a question regarding the behavior of consumers in certain conditions. He said to me (I paraphrase from memory): "my whole work is about finding the formula for this behavior. I know what they do, I have the graph drawn, and I am working to find the equation that would give me this graph." It sounded similar to the lifetime work of Einstein, seeking to identify the mathematical formula that would make credible what he already knew about the deep interdependency of mass, energy, and acceleration. That is because it was. Patrick Nasmyth was a renowned ocean physicist and an electrical engineer who worked on the development of radar during World War II. His wife, Mim (Mildred) Nasmyth, was a mathematician, and she helped him with mathematics when they both worked for the Defense Research Establishment Pacific, in British Columbia. So did Einstein's wife. This probably prompted Patrick Nasmyth to say, during his speech of retirement: "Now we have computers. When I started in this field we had girls." This is a statement that encapsulates, and masks behind a joke of the times, the reality of human calculators, a low status job that employed many people from the nineteenth century on, and the fact that computers were also developed to replace these "resource-consuming" humans with something more economically viable from a profit perspective. These positions were occupied mostly by women who were until recently forgotten; they are now celebrated for their role played in both the war effort as code breakers and subsequently in space missions with NASA. But, as probably many women who knew math do, Mim Nasmyth knew that materiality and mathematics mix in a problematic way. Her son, Guy Nasmyth, who is a System Thinking guy and a dear colleague to me, recalls her saying: "I didn't like applying math to anything physical because that interfered with the purity of the math." Most people in science know this, and there is little or no illusion regarding the relationship between physics/materiality and mathematics. A beautiful literary expression of this problematic relationship is Cormac McCarthy (2022) diptyque, *The Passenger and Stella Maris*, a story of the deep feelings with incestuous undertones between a young woman who is a mathematician and her physicist brother.

However, at the general public level, and in many realms of governmentality and administration, we encounter a very nineteenth-century positivist approach, equating quantification with predictability, certainty, and objectivity. The belief in question here that facts are expressed in numbers that speak the truth about phenomena turned into an epistemic requirement. It fed into two major directions: first, the tendency to transform everything into quantifiable units prone to produce data, and second, the propensity to ignore (or deem irrelevant) what seems to be not easily quantifiable or "just" qualitative data. This belief also makes any data-based product, from AI to research based on measurable parameters, an easy sale.

Two operations are crucial: making data and making sense of data. The first implies three major suboperations: the producing of data, the shaving of the "surplus of meaning" in order to generate quantifiable entries, aka data cleaning, and the production of new, secondary data based on this initial operation—with little or no material correspondent. This corresponds to the remaking of the material world into the mathematical—let us call it digital—world. Data cleaning is done by "farmers" who tend to be people being confronted daily with vast amounts of data under the form of tables—very similar to ledgers but in electronic format. The farmers need to evaluate and eliminate data that are "inaccurate," "redundant," "incomplete," and "irrelevant" (Stedman, 2022). The decision regarding what is relevant and accurate is done in relationship to the *specification of the algorithm needs and not in relationship with the environment from which the data was extracted.*

The process is also dependent on the worldviews of those who clean the data sets, as they are the ones who decide on the meaning of the "phrase" that needs to be encoded. An interesting account of Venezuelan data cleaners can be read in an article published in El Pais—a Spanish journal. The article explains how the economy of data cleaning relies on workers from lower-wage countries, such as Venezuela; aggregators of data subcontract their work and the workers are responsible for encoding and helping the training of decision-making algorithms: "Data taggers or annotators such as Fuentes provide information to computational models, so that they can make decisions. This includes everything from improving web searches to enabling more complex algorithms, like those needed for a self-driving car to work. "The AI system is watching and learning from what [these workers] do," explains Dr. Alberto Delgado, an AI expert at the National University of Colombia" (Medina, 2023).

These dependencies are most clear in social data sets, data produced by humans through their use of data-producing devices, from computers to smartphones. Our digital traces that is. The manuals and methods developed to minimize the impact of cultural categories on data sets can be compared with the manuals on double-entry ledgers, plus the cultural awareness gained. Take, for example, the article "Social Data: Biases, Methodological Pitfalls, and Ethical Boundaries" by Olteanu et al. (2019) in which the authors provide a thorough analysis of these methods and implications of bias introduced in the process. This allows the authors to show how volume does not supplement accuracy. In other words, big data is not necessarily a guarantee for accuracy, let alone certitude. They also list and explain sources of bias, including different cultural uses of the devices that produce data. However, the method proposed assumes that data is *collected*, not *produced.* No matter that in the very description of this collection, it is clear that we deal with a process of production resulting from interactions of tools and environment that cannot be free of bias. The article assumes that if we account for the biases through these methods, the data could be a perfect reflection of reality. In other words, if the process of cleaning the data is thorough, we will somehow obtain a spotless mirror of reality. This ignores the fact that those very spots *are part of the process and integrative to materiality;* they are pieces of reality that get eliminated in order for the data to match the research goal. If we hope to explain anything about (social) reality, we need not to eliminate the "spots" but to address the major cultural bias in question: the belief that precision means accuracy and that it can afford certainty. The mirage of objectivity is built through this bias.

This brings us to the second operation, making sense of data: the interpretation of the data as facts, as if they are direct, immediate indicators of material reality—a reality that is nonetheless a mathematical simulation. This process of sense making, like all other processes of making, is a cultural one that involves many people and that corresponds to a goal. In this case, the overarching goal implies the mathematical representation of an imaginary material world as if it has a real correspondent, as in Vermeer or Velazquez paintings, or as in Borges story about mapping recounted in the previous chapter. The cultural imperative of digitization is that of rendering the world perfectly intelligible in mathematical language. This, of course, is done by the multiplication *ad infinitum* of the data points—the granularity of data.

The digitized reality implies that what is taken into consideration from the material world are those elements that potentially can be data points. It also means that those that are not are to be disregarded. They do not *count*.

The digitized representation is a reconstruction of materiality using data points. What is missing are the possible spots on the mirror. For example, facial-recognition software extracts from the image elements deemed to be important for matching, eliminating all others. There is a vast amount of literature already that discusses the biases and failures of these systems. A step forward are the emotion recognition softwares that in a similar way generate readings of the emotional state based on assemblages of image data points that are somehow supposed to reflect emotional states. There are multiple layers of suppositions at play here, one that emotions are expressed on faces in similar ways, and that these expressions could be encoded and decoded through data points. I played with one of those emotion decoders, and what attracted my attention, beyond the expected errors, is the flatness of the representation (Fig. 5.3).

It brought me back to the transformations in drawings with the collapse of the vanishing point into punctum. With the idea of granularity of materiality as a sum of data points, every data point becomes a punctum, and the depth of the image disappears, what emerges is an image of a flat network that covers the three dimensional original. The same effect is visible in other representations that suggest digitization, of architecture, or of the world. The so-called instagramable faces—(self-)

Fig. 5.3 Emotion decoding—Iasi Culture Palace, Romania. The system reads this pose as "neutral." The tilting of the head is obviously not registered. Photo by the author

portraits using filters that erase wrinkles and "imperfections"—became desirable as reality, and people go through cosmetic operations to render the surface of their face smooth and flat. Psychological studies show that one of the important elements of communication and trust is the visibility of wrinkles in others. We trust others because we see that person's predominant expression through their wrinkles. Effacing them may contribute to the gradual loss of trust we have recently developed in our societies. Perfect faces, remodeled after digitally enhanced portraits, populate a trustless space. This is the space of the "Digital Republic" as Susskind (2022) calls the possible new data-driven political world.

With all its beauty and perfection, the mathematical system is infinite and yet bidimensional, and the binary system of digitization is even more so. Its effects are indeed retranslatable in three-dimensional materiality but what comes out is a flatter reality. After the creation of perspective, the emergence of dataism returns the world representations to flatness. Maybe the rise of flat earthers' beliefs is only an expression, a symptom that is, of the flattening of perception that our devices provide, from their material appearance—flat screens as the pinnacle of progress—to the breaking and schematic representations of three-dimensional spaces in bidimensional networks of data points as illustrated in Fig. 5.4.

The return to bidimensionality is both perceptive (screens, network designs) and in the practice of thinking about the world, politics included, in binary algorithmic terms. The rise of extremism in the past 15 years is attributed greatly to social media and the use of its attention-grabbing economy to polarize views.

Deeper than this, it is the very belief in the "modern fact" as objective, accurate representation of the world, and the possibility of reality's digitization opened by this principle that transformed the semantic complexity of the world into a syntactic binary expression of measurable elements. It inevitably led to social and political polarizations: if 0, then…, if 1 then… down the algorithmic road, with no

Fig. 5.4 Aya Sofia on a giant screen in Frankfurt International Airport—photo by the author

in-betweens. As Ruckenstein (2023) observes, reference to algorithms changes us[3] before we can catch up with it:

> We are living the metrics. This is where new patterns of experience begin to emerge, and people think differently about themselves and act accordingly. Algorithmic relations take hold of the everyday bit by bit, so that our recognition of what is happening often trails behind; we notice afterward how our behavior, or some aspect of our lives, has become modified by a specific metric technology. (Ruckenstein, 2023: 19)

The irony is that, unlike the people in my childhood joke, we are not aware anymore that "it matters how we say it," even if we utter a number. When the data expressing a joke met a new contextual reality in the train compartment, its meaning got lost. In the data-driven culture, *we lose the meaning of context,* intentionally or not. It does not matter anymore "how we say it" because we are culturally biased to listen to the *precise,* impersonal and *presumably accurate, objective, and certain* speech of the numbers. We desire certainty and manageability, and we find it in "the sanctity of data" (Bartoletti, 2020). Similarly, we eliminate context and react to associations and codified language, the soundbites that plague any current communication. No patience for the context. Social media codifies reality in decontextualized sound-*bytes* that the new algorithmic mind is processing in a simplified manner. Context is irrelevant, and syntax is all that matters: associations of signals that indignates, infuriates, and eventually makes people act on impulse—just like AI. In this context, we may as well remember a Goethe quote that never made it as a meme: "To act is easy. To think is hard. To act according to your thoughts is troublesome." Knowing all this, perhaps it is high time to get into the trouble of aligning acts and thoughtfulness.

The digital world has two major attributes: it is potentially infinite, and it is conceived as immortal—because apparently immaterial. While this chapter was mostly about the epistemic implications of "the modern fact," the next will tackle AI and governmentality, the fascination of subjectivity in the digital space, and the quest for immortality and its pitfalls.

References

Bartoletti, I. (2020). *An artificial revolution. On power, politics, and AI.* The Indigo Press.

Hobbes, T. (1982). *Leviathan*. Penguin Classics.

Medina, P. (2023, June 24). The Venezuelan ghost workers who are feeding artificial intelligence [article]. *El País*, USA Edition. Accessed November 6, 2023, from https://english.elpais.com/international/2023-06-24/the-venezuelan-ghost-workers-who-are-feeding-artificial-intelligence.html

McCarthy, C. (2022). *The passenger box set: The passenger, Stella Maris*. Knopf.

[3] The very title of this book was "Artificial? Naturally!" and it proved to be, in the publishers' view, too poor in keywords for the search algorithms to pick up.

Olteanu, A., Castillo, C., Diaz, F., & Kıcıman, E. (2019, July 11). Social data: Biases, methodological pitfalls, and ethical boundaries [article]. *Frontiers in Big Data, 2*(13). https://doi.org/10.3389/fdata.2019.00013

Pacioli, L. (1878 [1494]). *Tractatus de computis et scripturis [Treatise on accounts and records].* Camilla e Bertollero.

Poovey, M. (1998). *A history of the modern fact: Problems of knowledge in the sciences of wealth and society* (1st ed.). University of Chicago Press.

Rotman, B. (1993). *Signifying nothing: The semiotics of zero* (reprint edition). Stanford University Press.

Ruckenstein, M. (2023). *The feel of algorithms*. University of California Press.

Saint Augustin (2024). Confessions (Confesiuni). Humanitas Edition.

Stedman, C. (2022, January). *Data cleansing (data cleaning, data scrubbing).* [Blog post].

Susskind, J. (2022). *The digital republic. On freedom and democracy in the 21st century.* Bloomsbury Press.

Tsing, A. L. (2010). *Mushroom at the end of the world: On the possibility of life in capitalist ruins* (1st ed.). Princeton University Press.

Ullman, E. (1997). *Close to the machine: Technophilia and its discontents*. Searchlight Books.

Vidler, A. (1978). *The Scenes of the Street: Transformations in Ideal and Reality*, 1750–1871. in On Streets. editor Stanford Anderson, 29–113. Cambridge, Massachusetts, and London, England.: The MIT Press.

Wiener, A. (2020). *Uncanny valley: A memoir* (1st ed.). MCD/Farrar, Straus and Giroux.

Chapter 6

Survey and Reward

> "Do not forever with thy vailèd lids
> Seek for thy noble father in the dust.
> Thou know'st 'tis common: All that lives must die,
> Passing through nature to eternity."
> Queen Gertrude in Shakespeare, Hamlet, Act I, Scene 2
> Folger Library, (n.d. p.25)

The constant question remains: how do "nature" and "human nature" meet? And what does AI have to do with it? In the following pages, I will attempt to argue that our current understanding of "humanity" is the byproduct of the modern iteration of (metaphors of) nature through systems of surveillance, measurement, classification, and monitoring. We have relocated the newly found humanity between the solid rock of nature as a healing realm by excellence and the hard place of a nonetheless threatening one. Between Hobbes and Rousseau, every product, including us, comes with the combined promise of salvation and a threatening doom. We want our products to be ecological, bio, and we want to grow our own food, but we want to do it in an urban garden, away from the unleashed meteors. We also want our passage through nature to be smooth and eventually never-ending, an issue I am obsessively returning to, as it forms the basis of our separation from it. We claim we want a return to nature but a nature that is morally righteous and cleansed of its "wild side." But what is "wild" and where does it reside?

My great-grandmother taught me about the world through stories. Her own life story spanned a century, covering most of the 20th one as an adult, but also stories she had heard in her childhood at the end of the nineteenth century in her small village in the middle of Transylvania, at the time part of the Austro-Hungarian Empire. Among the stories that stayed with me is that of a festive table and its aftermath. Warning: the stories of the time and the way great grandmothers tell them may sound, and are, raw and cruel. They are also parables full of layered metaphors and meanings. This particular one goes like this:

A. Balasescu, *Climate Change in the Age of Artificial Intelligence*, Artificial Intelligence, Simulation and Society,
https://doi.org/10.1007/978-3-031-90042-6_6

At a large gathering around a copious meal—the occasion matters less, although I guess it might have been the commemoration of someone's passing—people had lively conversations and exchanges in a relaxed manner, affording the luxury of forgetfulness of their daily lives. Somewhere around the table, a man is yawning. Not far across from him, a young woman yawns, too. Her husband observed the sequence of microevents. He said nothing at the moment, but he is convinced that this was a secret sign between the two. His (jealous) mind started to devise a plan. A few days later he invited his wife for a hike in the neighboring forested hills. Carefully hidden in his backpack, there was a rope. As they walked deeper into the forest, his wife was intrigued by his demeanor. He was looking attentively at the crowns of the trees, approaching sometimes one trunk or another and moving along. At some point, exasperated by her husband's undivided attention to trees, she exclaimed: "Why are you walking from tree to tree, like the yawning from person to person?!" (in Romanian it rhymes: "Ce umbli din pom in pom, precum căscatul din om in om?!") He stopped, startled, and asked: "What do you mean?" His wife explains: "Haven't you noticed that every time someone yawns, someone else will also do it, and the yawn can go around without anyone wanting it?" The young husband suddenly felt ashamed. He confesses his dark thoughts to his wife: "I was looking for a branch strong enough to hang you… You know when we had that feast? You yawned once and I thought you were making signs with so and so… I was overwhelmed with jealousy… please forgive me." She does, and they return home and to their lives.

I am not sure why she told me this story when I was only 9–10 years old. Remembering it now, and in the key of this book, a multilayered way of interpreting becomes obvious: first, it is a story about the pitfalls of jealousy. It is also a story of patriarchal domination as a system of surveillance and punishment. In this story, the husband is implicitly invested with the power over death specific to the juridicolegal system in Foucault's terminology. In this type of system, the French philosopher explains, power is expressed as the possibility to decide the type and the moment of punishment/death over its subject. Once the subject commits an act considered a crime, the monarch/the patriarch exercises power through punishment. Previous to the act itself, the exercise of power is rather absent.

However, the story above also contains elements specific to the surveillance system, developed in the modern period, and the very creator of modernity, according to the same author, in which power exercises itself continuously on its subject, simultaneously constructing it. The exercise of surveillance is perhaps less interested in what someone, a preexistent subject, does but in how its actions could be both manners of shaping, and signs of, the subject in becoming. Those signs are traces of the possible future, and at the same time. they form the basis for current actions that bring the subject into being. In Foucault's terms, this corresponds to "power over life," in the sense that the actions during the entire life are to be surveilled, regulated, controlled, and corrected, and these combined actions of power create the subject in the continuous process of subjectivation.

In *Discipline and Punish*, the book that lays out Foucault's (1995 [1979]) theory of power that dominated the second half of the past century and continues to be relevant today, with its limitations and critiques, Foucault undertakes an archeology of power and identifies an important moment of transition in the way in which the Western sociopolitical system organized power: the shift from power over death to power over life. Politically, he roughly overlaps the power over death system with the pre-French Revolution monarchy and calls it the juridicolegal system. The passage to the system of surveillance and punishment (the original title of the book in French, *Surveiller et Punir*) is best illustrated by a prison project designed by Bentham, called the Panopticon. The principle was as simple as it promised to be efficient: a circular building on a few levels would have at its empty center a control tower. The individual cells of the incarcerated would face the tower through a transparent grilled wall/gate. The presence of one guard in the tower would be sufficient to survey the entire complex. What is better, the prisoner would never know if the guardian is present in the tower or not, the relation of transparency not being reciprocal. The subject of this power is the surveilled subject: "He is seen, but he does not see; he is an object of information, never a subject in communication," writes Foucault. The one subtlety is that the subject itself is indeed not an a priori part of the communication, but it becomes integral to it inasmuch as it generates the information that in turn constantly creates the subject itself.

The convicted would thus be compelled to permanently behave well, being under a potentially permanent state of surveillance. The combination of surveillance and the looming potential disciplinary retaliation inflicted by the system would be morally effective, monetarily efficient, and last but not least healthy: "Morals reformed, health preserved, industry invigorated, instruction diffuse, public burthens lightened. Economy seated, as it were, upon a rock. The gordian knot of the Poor-Laws are not cut, but untied. All by a simple idea in architecture"—Jeremy Bentham (2011), *The Panopticon Writings*.

Bentham's prison was never built in the UK, but the model was exported to the colonies. In Cuba, Vietnam, and the United States, prison complexes followed this blueprint. In Vietnam, they were used to imprison political dissidents, in Cuba they lay in ruin, and in the United States, the last roundhouse prison was closed in 2016, in Illinois. The architectural concept, although not necessarily in its round form, is still in use in the Twin Tower Jail in Los Angeles.

Foucault's argument goes beyond the prisons' walls, affirming that this is a system of thinking that pervades large swaths of the current social organization, combining surveillance and self-government.

Of course, the juridicolegal system of the "ancien régime" did not completely disappear but got buried under the new system of governmentality. Punishment still exists but away from the open, and disciplining through surveillance is favored. This system is at the foundation of the principles of governmentality, meaning simply that power is distributed in the society and is exercised through different means, by a variety of institutions that are not necessarily linked to the state. Corporations, schools, and financial systems are all part of the governmentality system, while the disciplined subject under constant surveillance is not (necessarily) repressed but mostly produced in this intertwined web of relations. Power is not repressive but productive. The subject of this web of power, the individual, is not preexistent and repressed by it, but it is simultaneously its creation.

The theory of governmentality was subsequently developed and refined by various authors such as Rabinow, Castel (1991), and Agamben in the 1990s and was particularly used to put under the question mark the idea of the "liberal subject" in Locke's term. The free-willing individual, the preexistent subject who moves in the field of forces that constitutes the society, is perhaps no more, if it ever was. Indeed, the subject is the product of these very forces and structures, as Maurer (1999) shows in his study of the insurance system (another practice of governmentality):

> Traces of the future help manage risk and control the unpredictability of temporality. The Lockean and Hegelian subject of property takes a back seat to a system of statuses based on one's investment in that temporality, as ownership itself evaporates into risk profiles. (Maurer, 1999: 67)

In the article *Forget Locke? From proprietor to Risk-Bearer in New-Logics of Finance*, the author identifies four historical stages in which the conceptualization of risk had successively changed, and these stages represent as many shifts in governmentality. Maurer sketches out the temporal inversion in the logic of insurance: risk is foreseen, calculated, and predicted using heterogeneous factors that constitute "traces of the future."

The field of insurance was a fertile ground to study (and test) governmentality, because the practice of insurance perfectly lays out the disciplining of the future. In the early 1990s, another author, Daniel Deferts, explains how insurance represents the perfect expression of governmentality and power over life that eliminates the haphazard of death and implicitly its limitations: "For death as a happy stroke of fortune which liquidates a debt, [the system] substitutes a plenitude of life determined by sex, age, hygiene, genealogy and family environment" (Deferts, 1991: 218). "[I]nsurance is like a diagram, a figure of social organization" or "a *generalizable technology* for rationalizing societies." (ibid., 215), and the insured, the entity that populates this diagram—that is each of us—is constituted of the classificatory parameters of their life. The individual is no longer subjected to liberal law but to its own calculated potentialities (both of risk and of success). It does not take much effort of imagination to understand what happens when AI systems are introduced in the insurance and finance business.

There is an already vast body of literature that shows the impact of AI in the insurance industry that can be summarized as follows: instead of redistributing risk to the social body, the automated system individualizes risks and penalizes already disenfranchised categories, based on proxy determinants, meaning factors that are attached to one's current life situation and that are not directly causal but loosely correlated with an increased risk. For example, what if someone lives in a low income neighborhood? The system may be programmed to process this as a risk factor and thus increase the insurance premium or decrease the chances for a loan. That is, the postal code becomes a factor in the insurance premium or in the likelihood of getting a loan for that matter. The decision is based on a measurement that has little or nothing to do with one's individual situation but a regularity that may have been statistically observed in the past, and then it is attached as a sentence to an individual's current and possible future becoming. The algorithmic loop creates a model and reinforces it upon the reality, weighing out whatever does not match. This type of approach perpetuates the status quo, as in fact the uncertainty of *becoming* is not in any way computed in the system nor possible to be integrated. This creates not less than an automated discrimination. As an illustration, in a very recent article in *The Atlantic*, Brown (2024) shows how insurance companies currently refuse to insure patients with genetic preconditions, setting the tone for "genetic discrimination."

The theory of surveillance is Foucault's most well known, and it led to the current consensus that we live in the global age of surveillance capitalism that takes a variety of forms and names. Its popularization also led to the stranger views in conspiracy theories of a diversity of groups, genre QAnon, who gave it a literal interpretation and postulated that "they (who have the power) are out to take us (who don't) by any technological means available." What these conspiracy theories fail to understand is that no one "has the power" because power is not an object or a quality to be possessed, but it is an emergent complex phenomenon that creates its own subjects. Power is of many types, and it does not belong to anyone, but it both emerges from and configures a field along with its subjects. In other words, you can always turn the camera (this symbol of surveillance and generalized panopticon), which is precisely what happens in many circumstances, including when QAnon adepts use social media and their algorithms to spread their theories.

Most recently the idea of surveillance society has been brought back to the fore by Shoshana Zuboff in her book *The Age of Surveillance Capitalism* (2019). The author makes a strong case about how data societies organize knowledge, surveillance, and ultimately life in a system concentrated on the single goal of profit making. As in Bentham's writing, the author observes that the capitalist consumer societies by means of digital surveillance and subtle behavioral modifications reform the individual and transform it into a predictable entity, while markets are selling predictions of this future behavior. Simply put, when a social media site sells a targeted publicity space, the bet is that in the future the target will buy the product, and the prediction of this future action is based on the data already accumulated about the target. The media company sells the probability of the target behaving in a certain way, the claim being that the target is already prone to do so, and only needs a nudge in that direction. However, the accuracy of these predictions is based on vast amounts of data, obtained through our own usage of the apparently free system.

The dark side of this approach transformed by the economy of attention-grabbing has been widely discussed, particularly regarding two instances: the "prediction" of individual and social behavior and delinquency and more recently the prediction and reinforcement of political views. While the first concerns continuously lead to further discrimination, biased policing, and wrongful convictions, the second has undermined the political process based on debate and dialogue.

The reinforced learning system proper to algorithms, based on the assignment of increased value of truth to sentences with more frequency of appearing, spilled into the society through social media channels and through the application of decision-making algorithms indiscriminately into realms that may not be proper for it, such as justice or the individual financial systems of loans. The algorithmic system does not understand if something is true or not; however its truth is given by the frequency of clicks on a certain topic. Every variation from whatever is reinforced by frequency is weighed out and disqualified as false; therefore in more ways than one, the result is reinforced bias (as discussed in previous chapters).

Socially, this translates into biased decisions in the cases in which the algorithms used for decision-making remodels reality using proxy parameters (i.e., postal code or online behavior as predictive of actions). Gender, race, age, residence, or online browsing history are reinterpreted through a model that assigns a predicted future that more often than not is self-fulfilling. If the postal code disqualifies one from getting a bank loan, their chances of growth are reduced, and this reinforces the "algorithm's view" that people coming from that respective postal code are not "worthy" of loans. The same thing happens with racial, gender, and name biases in hiring processes, for example. Certainly the AI ethics field, greatly developed in the past few years, is trying to understand and mitigate these occurrences. The consequences of treating social reality as a model have been very well discussed by Cathy O'Neal in her *Weapons of Math Destruction*:

> When automatic systems sift through our data to size up for an e-score, they naturally project the past into the future. As we saw in recidivism sentencing models and predatory loan algorithms, the poor are expected to remain poor forever and are treated accordingly—denied opportunities, jailed more often, and gouged for services and loans. It's inexorable, often hidden and beyond appeal, and unfair. (O'Neil, 2017: 155)

The core of a surveillance, data-driven society is indeed analogous to the core of the insurance industry as described above: controlling the uncertainty of temporality—regardless of what the future may hold. Massive data analysis can give the illusion of control. It is as if the entire architecture of society would purchase an insurance policy.

Politically, by feeding a confirmation bias in the population, it attracts toward more and more extreme views that reinforce the data-founded truth. Those exposed to only one set of information, fed to them by the algorithms that confirm biases, will end up accepting it as the truth and start behaving algorithmically: everything that does not match it is discarded, weighed out, or unworthy to engage with. This led to an automation of social and political behavior, in which the identification with a "virtual truth" obtained by reinforcement, such as "the world is flat" for example, is more important than the debate about the value of that truth possibly verifiable outside the system that created it in the first place. We are currently experiencing the consequences of this mechanism that recreates the political world on binary oppositions, more often than not irreconcilable: the shrinking space of debate, the extremization of political discourses, and the a priori dismissal of the other because of its otherness, all these behaviors crept into the public space and restructured it. Public space did not disappear, but it is emptied of interactions and filled exclusively with self-affirmations, like a generalized Instagram or TikTok. This leaves few possibilities for "the other" to become a follower (as in identifying with, becoming the same), to retreat in its own vehement self-affirmation and enter a contest of gaining followers—based not on debate but on a more vehement affirmation of the self—or to simply leave. The result is Brexit-like.

The motivations and the role of the subject, the individual, as it is perpetually created within the system, are somehow different from the classic view that associates surveillance with punishment. In many of the instances of surveillance capitalism, instant gratification or promised reward replaces punishment. The data needed for the large-scale AI models are generated by us, "the users," in exchange for a promise of the future: a happier, healthier one. Through technological assemblages that promise a variety of fulfilling desires, we consent in producing and sharing the data the system needs. I will not comment here on the ethics of this but on its triggers and mechanics. The promised happiness and fulfilment are defined through the very interaction and use of the technology. The number of likes, the rewards for consumption—translated usually into more consumption—and the statistics about our bodily performance generated by our health and well-being apps are nudging "the user" toward the narrow definitions of happiness as proposed by the data gathering system itself. These are accompanied by predefined behaviors that indicate a performance increase in the system: we get X% reduction in future purchases if we do this or that, we get a badge for more engagement, or we receive a pat on the shoulder and the promise of a longer life if our caloric intake and our walking habits "improve."

Anthropologists from JP Warnier (1999) to Tim Ingold (2013) emphasized that we think together and together with our tools, not separated from them as "individuals." We are part of hybrid systems of thinking in which agency is distributed, as Latour would put it, or of assemblages and affordances in Hayles' terms. It is the hybrid between the individual and the technology it uses, between the human and the mobile phone, for example, that act in the world, all the while creating the possibility and the field of those actions. Agency is distributed and exercised as such, and the assembly is transforming together.

Perhaps this is the profound sense in which surveillance capitalism changes human behavior. Not because it corrects that behavior of a preexistent subject through a system of punishments or encourages it toward a desired direction with a series of rewards (a very behaviorist view by the way), but because the very behavior is shaped within this interaction and tends toward a common denominator between the algorithmic binary and its translation in human minds and communities. Zuboff is right as long as we think of the individual as preexistent to the system of surveillance, an individual whose behavior is corrupted and modified by this capillary system. I propose a deeper interpretation: the profound change is not of an ideal individual that "falls" from a supposed "state of grace of control and absolute agency" into the "hell of manipulation." This is precisely the narrow understanding that conspiracy theories assign to surveillance societies of our time, rushing into conclusions about malevolent agents that exercise the power of control for some nefarious preexisting goals.

This is the risk of superficial readings of human agency and what surveillance capitalism means. A rushed read of Zuboff's book may bring someone to this conclusion. If we are to consider the individual separated from the assembly, the profound change in the agency of the individual is that it functions less in complexity and more on an algorithmic rule, becoming thus intelligible for the algorithm itself, that reinforces this *modus operandi* in its turn. In other words, we become more like the machines we create, through the process of attributing to them human ideal characteristics (objectivity and rationality, both analyzed in the previous chapter) and we then try to emulate them.

Perhaps the danger is not the taking over of the world by machines (to great degree this already happened in the process of industrialization) but the taking over of our mind by the algorithmic method of thinking that eliminates any form of effect of biology, life, emotions, and experiential unicity. This elimination, this effacing of affection for the world, is regarded as the hoped-for future; the promise and aspiration of AI by a large part of the champions of AI and certainly is a characteristic of the Silicon Valley culture. Why? Simply because, in Hannah Arendt's words, biology (and its materiality) is a reminder of death: "The mortality of men lies in the fact that individual life, with a recognizable life story from birth to death, rises out of biological life" (Arendt, 2018 [1958]: 19).

"The world alienation," discussed in Chap. 1 from the perspective of the escape from materiality, comprises also the deeper desire to escape one's own material conditions of existence. Among the multilayered implications analyzed by the author, the most prominent is that desire to "escape" the confinement of the Earth that became possible with the technological progress and first space travels, in a movement that marks the saving from the indeterminacy of biological life. As if by eliminating biology, it would be possible to eliminate death. Death itself, in an AI centric world, is an inconvenience, as Kneese (2023) observes. It is a reminder of the false promise of technosolutionism, the ultimate glitch in an otherwise "perfect" system.

In a Reformist worldview, it is not death per se that is threatening, but life itself is a nuisance since it is filled with the uncertainty of salvation, and the self needs constant reassurance. The exit from this indeterminacy can be devised in various ways. Traditionally, money constituted a good enough proxy for confirmation—the thesis upon which Max Weber's theory of capitalism is based: the pursuit for profit comes from the uncertainty of being saved, financial success being the closest confirmation one can aspire to (Weber, 2001 [1904]). The two other inventions, argued Arendt, enabled different promises of the same exit: through modeling life on Earth on universal laws and effectively through space travel. The variety of paths for the confirmation of a predestined salvation is expressed in one concept: the control of variations through mathematical universal laws. Money, after all, is also a mathematical remodeling of human worthiness.

However this will only be possible in this secular life by separating from the indeterminacies of life itself—and the world. This would give certainty, confirm the saved status of the chosen ones, and eliminate variations: control of variables, elimination of uncertainty, and denial of death, since the final moment of transcendence is moved to the point of origin—the birth of the predestined soul for which life is inconvenient and death is irrelevant. The only acceptable form of life is the possibly manageable and ideally rationally administered, codified one. The obsession with, and the particular understanding of the DNA code, is a clear indicator of this view. Paul Rabinow's research in the failed merger of two pharmaceutical companies, a French and an American one, showed how important is the cultural understanding of life. In *French DNA. Trouble in Purgatory* (1999), he convincingly argues that the failure is due to the different if not opposed view on life: pure life, *zoon,* the quality of being alive with which the puritan American culture operates versus *bios*, meaning life plus all its qualities, uncertainties, and becomings, with its biography so to speak, proper to the understanding of a catholic based culture. For the former, DNA is the key code that prescribes, predestines, and thus renders irrelevant the latter. Nature (the mathematical one) is predestined and the only relevant one. Bios, the nurture and its impact on the becoming, is unimportant in this view because becoming is irrelevant for predestination. One can always return to the preexistent code for answers. Therefore, as Ezra Klein puts it in his conversation with Erik Davis, it is crucially important to understand the culture of people who create AI, particularly the weirdness of Silicon Valley and its puritan undertones.

The moving of the individual destinies into a data-driven model based on surveillance is ultimately yet another modality of exit from the messiness of life and a disciplining of biological-based society into a predictability realm that eliminates the puritan insecurity of predestination—a Leibnizian dream of mathematization come true. There are a few pertinent analyses of the genealogy of thinking AI and large data modelings, linking them to a mainly protestant and predominantly male worldview. Both Meredith Broussard (2018) and Franklin Foer (2017) trace the ancestry of the current vision of the tech world in this direction. Broussard points out at the entire vocabulary of Silicon Valley imbued with metaphors of salvation and rapture into the "machine," guided by gurus, prophets, and angel investors towered over by the figure of Ray Kurzweil.

The most comprehensive book to date on the intimate and intricate tapestry woven from religion, philosophical thought, and dreams of an artificial intelligence operating in a mathematical universe is *God, Human, Animal, Machine*. The author, Meghan O'Gieblyn (2021), shows how the aspirations embedded and embodied in automated systems that offer mathematical models of reality and act upon reality itself as if those models are true are deeply religious and of a certain type. She argues that God was not killed by secularization but that it keeps coming back through new eschatological forms that promise the resurrection. Transhumanism and technological singularity are terms for technological salvation, the exit from the imperfect world of biology into the perfect (and presumed perfectly objective and rational) one of computation. In this view, the moment of reaching artificial general intelligence (AGI) should ideally correspond with the merging of human life with the super intelligence.

The language of Earthly salvation directly appears in healthcare and well-being when they meet technology, like in the figure below. Salvation is around the corner once human imperfection is eliminated: the biological one from within (the

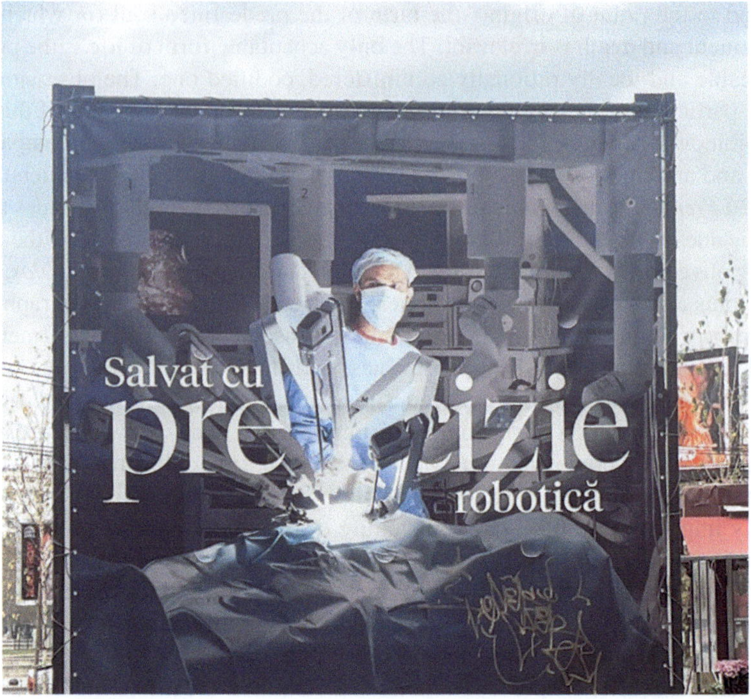

Fig. 6.1 "Saved with Robotic Precision"—street publicity for a private clinic in Bucharest, Romania

sickness), and the existential, presumably error-bound one (the surgeon's hand) from outside. (Fig. 6.1).

As in any eschatology, the fate of those "left behind" by this process is dire. A patient of the new technologies, herself a retired medical doctor, recounted the shock she had when she expressed her caution regarding her own frailty. She informed her doctor that she is somehow afraid that her weakened body would not resist a new procedure proposed to her. Her curing MD answered: "Well, madame, it is like this, there is nothing we can do about it. Some patients resist, others…" Of course, AI use in healthcare has tremendous benefits. But the precedence of the process optimization over the well-being of the patients is a tendency that indicates the "reverence to the machine" and that needs to be kept in check with discernment.

The current use of AI in domains such as finance, justice, or human resources operates already (with) its own left-behind categories, and the impact on this is documented and visible on human destinies. Even our daily use of apps from search engines to well-being tools feed into the system of creation, and reinforcement, of predestined identities. Books such as *Automating Inequality: How High-Tech Tools Profile, Police, and Punish the Poor* (Eubanks, 2018) detail the mechanisms of the data-based confinement of life into the cage built with the bars of the traces of an already predetermined future.

The data-based predictions are sets of self-fulfilling prophecies in a vast field of variations on the same, a universe of the repetition of the same (identity), regardless of how big the data set or how sophisticated, opaque, and apparently surprising the algorithm might be. Governing the algorithmic *zoon* in the space of data surveillance is an administrative paradise that operates (with) the minimization of variations, since variations are weighed and reassigned to certitudes. Governmentality based on surveillance is best adapted to identitary spaces and definitive identification of subjects.

The Italian philosopher of law Giorgio Agamben (1998) makes the clear argument that the neoliberal administrative forms of government today are less preoccupied with the politics of living together and more geared toward governing pure life, *the zoon.* This makes the AI systems all the more relevant for the apparatus of power, through their capacity of using data generated by our biographies, our *bios*, in order to recode it into some form of pure life, intelligible for algorithms. The only way out is refusal to participate.[1] but this would mean death from the data space. The finitude of the biological body is the reminder of *bios*, a threat to the system, an easy way out. By carefully listening to the way death is described in various instances, we can hear its meaning. In the field of insurance, we describe it as "loss of life." In the circles of West coast researchers in longevity, we hear that aging is a disease and death is unnatural. That is, death is our "wild side." We can cure it. In the debt-based finance system, death is almost an illicit avoidance of paying one's dues. Suicide is frowned upon by both insurance companies and religions: it is the avoidance of paying the moral debt of the original sin (Graeber, 2012). Death is, most of all and in all instances above, unnamed, resonating perhaps with the way in which the 13th Tarot Card was unnamed in divination circles. Death, within this system, is immoral. Biology, and by extension, the wilderness, is immoral, and the body is a reminder of this immorality, not because it is sinful but because it is mortal. The code (including the DNA) is not. Mortality as the ultimate form of immorality is in fact the undesired aspect of "human nature" that the surveillance system promises to weigh out of our lives when moving them into the perfect algorithmic space. The affordability of molecular medicine is the possibility to return to the code, to act upon it, and eventually modify it in order to obtain the desired salvation (perfect health, immortality) and of course only if one can financially afford it.

But what is the space of action-performance for data subjectivities produced by the diffuse surveillance apparatus? Let us further the conversation with Foucault's insights into the genealogy of governmentality and the way it relates to surveillance. In his less popular outside academic circle courses at the end of the 1970s and early 1980s, the French philosopher details a second important historical rupture that, building on surveillance, created the decisive paradigm and the scope of governmentality: the securitization. While surveillance is necessary for identifying, disciplining, and "correcting" the (individual) subject, securitization refers to the entire ensemble of society and the necessity of its functioning:

[1] See Feminist Data Manifest. [No Manifestno.com] (Website). —manifestno.com

> [...]it is enough to see the legislative system and the system of disciplinary obligations as part of the modern mechanisms of security to understand that there is no succession:[2] law, then discipline, then securitization; but that security is a certain modality of adding, of making work not only its own mechanisms, but also the old joints of law and discipline, (Foucault, 2004:12, my translation)

Let us get back for a moment to my great-grandma story. In the young husband's imagination, the sequential yawning is not a proof of the existence of mirror neurons but a sign of possible actions that condemns the subject to a trajectory of becoming, namely, his wife to hanging.

The passing of the yawning is a natural occurrence but something that, in the husband's mindset, requires control. The yawning is thus a trace of the future, a future based on inferences through proxies. It is not about the yawning itself as a contextual fact but about the existence "of a signaling" of some kind. The yawning is a data point that is abstracted and made equal with any other gesture of illicit communication between the two future adulterers. This is what data does—but let us not anticipate. In the young husband's abstract and decontextualized view, the traveling yawn signals the transgression of another type of natural law: the morally righteous naturalized one of gender-based surveillance, control, and application of that very law.

However, the surveillance here, to Foucault's point, extends from the subject of power to the entire context, and it is applied in order to secure the future. The important element that comes into being in the paradigm of securitization is "the environment"—that is the ensemble of the conditions in which the technologies of power are effective.

This environment itself is imagined to be ruled by laws; it is intelligible and thus controllable. Its emergence, in Foucault's view, is marked by the apparition of categories such as "population" and the effective way in which the environment functions is "the disciplinary normalization," that is, the standardization and elimination of "aleas." The environment, in this view, emerges from the application of these very technologies; it is the reconstructed reality in terms that renders it intelligible to power.

> [...] securitization will seek to organize an environment in function of the events or the possible series of events that it will need to regularize in a multivalent and transformable framework. [it] refers thus to a series of possible events, it sends us to temporality and randomness, which it will need to circumscribe into a given space. (Foucault, idem.)

The invention of the environment is linked to social space and human behavior, and the environment itself needs to be circumscribed in order to function *as an environment.* The relatively enclosed urban spaces were the frame of the potentially securitized environment: a lab condition in the real world, where parameters can be created, identified, measured, and controlled for some ultimate objectives. It is relevant to emphasize that the environment was not originally conceived as a mode of approaching the "natural" realm but the "artificial," human-made and populated spaces. "Wilderness" was still too vast in the imaginary of *discovery and*

[2] Meaning that one system did not replace completely the other.

exploration that pervaded the eighteenth and nineteenth centuries and formed the backbone of the colonial imaginary, when the writings about the management of the environment emerged. The world was not yet circumscribed.

Gradually, this concept of environment extended to the wilderness by linking itself with the invention of "nature," through a unification of laws, and the moralization of nature. As we previously saw, the natural realm was slowly becoming measurable and understandable in terms of regularities of occurrences bounded in the spaces and geographies of climate. The need here is to remind ourselves of von Humboldt's observations on the correspondences between climate regions and the height of mountains and between latitude and altitude. Through the operation of mapping, measurement, and quantification, wilderness gradually becomes "nature." The unboundedness of the world could thus be bounded in climate areas, meteorological similar territories, prone to later become "natural environments," governed by "natural laws," and thus manageable. As Hannah Arendt observed in *The Human Condition*, the conquest of the American continents marked in the Western imagination the beginning of the boundedness of the world—and thus constituted the first step toward the possibility of extending the notion of environment to a global scale. The final step that certified the boundedness of the world is of course the first photo of the Earth, the blue marble, from space. It was the turning of the telescope toward ourselves, from the external Archimedean point outside our world, the birth of the idea of a global environment, and the moment that validated in everyone's eyes the possibility of thinking of our planet as an environment, singular. From Hannah Arendt to Donna Harraway and from Kurt Vonnegut to Lovelock, many thinkers commented on the implication of this change of perspective. It is worth mentioning that for Alexander von Humboldt himself, nature was not an environment, but rather it was an open system characterized by deep interconnectedness.

For our purpose, let us remember that the idea of environment allowed the solidification of the view of the world as a possible object, an immense bounded system that eventually can be modeled. Within it, the seed of the idea of global conservation is also present. I will provide a longer conversation on conservation in the next chapter, but let us think here of conservation as the elimination of changes. Just like in the case of the individual person that reaches for immortality, one of the conservation underlying ideas is that the planetary system may be conserved in the current form or even improved. The assumption is of course that the system is a closed one (bounded). And if not improved, perhaps cloned?

In his series of talks *The Vital Illusion* given in 1999 at UC Irvine, Jean Baudrillard offered his unique view on the decoupling from what he calls "the real," bringing into discussion the project Biosphere 2, the miniature cloning of Earth's ecosystem in Arizona. Five climatic zones are reproduced and surveilled in "ideal" conditions, having perhaps as scope the idea of reproducing them elsewhere.

> The specter that haunts genetic manipulation is the genetic ideal, a perfect model obtained through the elimination of all negative traits. In the experimental prototype Biosphere [2], for example, we find no viruses, no germs, no scorpions . . . and no sexual reproduction. Everything in Biosphere [2] has been purified, immunized—immortalized—through transparency, disincarnation, and prophylactic disinfection.
>
> (Baudrillard & Witwer, 2000: 20)

Biosphere 2 is the perfectly conserved environment in its current form but also one in which death does not exist—individual death is eliminated together with our possible extinction as a species. The world alienation is completed by building a perfectly natural world without its uncontrollable "wild" side, a world that can feed the recently revived Mars colonizing dreams by the same Silicon Valley AI elite. As a young PhD student at UC Irvine, I had the pleasure and privilege to be in the amphitheater for Baudrillard's talks. During the Q&A session, someone asked Baudrillard "what is the place of alienation in this world you describe?" The French philosopher looked above the thick frame of his signature glasses, smiled, and answered in his thickly accented English: "Alienation?!... ahaha! Good old times, good old times…"—of course, because in those "good old times" there was still an existing referent, a world to be alienated from.

The AI survey of the planet and its mapping as part of the solution for climate change is the project Biosphere 0. It clones the planet into a dataset, in an always already simulated reality on which afterward we can act, imagining that we act on the entirety of our world. The unification of the planet into a singular environment is a unification within the realm of control and influence. This is not the undoing of the world alienation, not the reunification of the human species with the world, as in replacing (and eventually replacing) the human in the biosphere but bringing the biosphere in the secular human realm, to be further controlled and thoroughly exploited.

Human and biosphere are reunited indeed but as data generator entities. In order to understand this, we need to deeper understand that the enhanced surveillance and the system of rewards designed to both generate data and domesticate individual behavior is a reflection of our relationship with the concept of nature—both human and nonhuman.

If nature was historically a product of observation and classification, our aspiration to build a perfect society brings politics back to nature as we define it. Unfortunately, human classifications are not neutral. On the contrary, bias, inequalities, and hierarchy are embedded in them.

As nature appears surveilled, measured, and possibly saved from climate change through networks of sensors generating data, so are we regenerated in systems of surveillance in order to be saved from our own nature—and ultimately rendered immortal. The argument here is that the system of surveillance and reward in the population governance of today has as its ultimate goal, the eradication of finitude. In the process, we recreate and operate with highly problematic classifications of human beings, and we renaturalize it with the help of technology.

On the one hand, we celebrate the return to nature as a solution to the general malaise of our modern lifestyle. On the other hand, we feel the need to survey and control any aspect of our behavior that would drag us back to our Hobbesian "worse nature," to the "wild side." But what frightens us about our "wild side"? What is the beast that needs to be tamed?

The immediate answer would be the "devious" behavior toward which some pretend we are naturally inclined. From the fear of the "letting go" of our bodies that would naturally gain weight or decay, to letting go of our supposed predatory, individualistic instincts that would push many to become thugs, the fight against this nature is omnipresent. Our devices survey our behavior, internally and externally, in the same manner we survey our natural environment (with justified anxiety) and fret over the effects of climate change. We seem to think that all these systems will keep everyone within the norm and secure the perpetuation of life within current parameters.

The alternative answer is that this deviation might only be a symptom, an indicator of the possibility of death. The deviation from the norm of capitalism is death itself: death as the finitude of the human body, death as planetary extinction caused by climate change, and death as full stop in the complex system of debt and finance. We fear unbound wilderness because it is a constant reminder of our own ephemerality, so we try to eliminate its traces by transforming it into bounded nature.

Finally, my great-grandmother's story is also a story of system self-correcting. Faced with new evidence, the husband questions his own actions and decisions and changes action in its course. Perhaps this is what my great-grandmother intended to transmit to me. There is always time to correct one's views, in an open dynamic system. This is also what Cathy O'Neil had in mind when she talked about the possibility of self-correction: "[...] the world is dominated by automatic systems chomping away at our error-ridden dossiers. They urgently require the context, common sense, and fairness that only humans can provide" (O'Neil, 2017: 155). I interrupt here her train of thought to accentuate "humans" and their defining quality: mortality. Only humans with the consciousness of their mortality can and hopefully will provide the understanding of the context as an open system in which fairness and justice are possible because emotions are allowed in. She goes on: "However, if we leave this issue to the marketplace, which prizes efficiency, growth, and cashflow (while tolerating a certain degree of errors), meddling humans will be instructed to stay clear from the machinery" (O'Neil, Idem). At the same time, ethics may altogether disappear in the new political space we construct for humans. Simply because in the world restricted by technologically determined choices, the question will not be anymore how "to do the right thing for the right reason." "It will be harder to teach ethics to our children when the world makes them ask not what is *right*, but what is *possible*" (Susskind, 2022: 82).

What if we, with our algorithmic modeling, are about to close the system? The danger is real as long as we invest AI with pure rationality, infallibility, and imagine it as the perfect system of world building. What if we are just about ready, in our desire for control of life and elimination of death, to hang the world from the illusory solid AI branch in the tree of quantified rationalism?

The next couple of chapters will look away from it, in the direction of other than modern ways of thinking, doing, and being and other-than-human forms of existence.

References

Arendt, H. (2018 [1958]). The human condition (1st ed.). The University of Chicago Press.

Agamben, G. (1998). *Homo Sacer. Sovereign power and bare life*. Stanford University Press.

Baudrillard, J., & Witwer, J. (2000). *The vital illusion*. Columbia University Press. [ProQuest Ebook Central doi unavailable].

Bentham, J. (2011). *The Panopticon writings*. Verso Books.
Broussard, M. (2018). *Artificial unintelligence. How computers misunderstand the world.* MIT Press.
Brown, K. V. (2024, November 12). Genetic discrimination is coming for us all. *The Atlantic.* https://www.theatlantic.com/health/archive/2024/11/dna-genetic-discrimination-insurance-privacy/680626/ (Website)
Castel, R. (1991). From dangerousness to risk. In C. Gordon & P. Miller (Eds.), *The Foucault effect: Studies in governmentality* (pp. 281–299). University of Chicago Press.
Deferts, D. (1991). *Dits et écrits: 1954-1988, I* (2nd ed.). Gallimard.
Eubanks, V. (2018). *Automating inequality: How high-tech tools profile, police, and punish the poor* (1st ed.). St. Martin's Press.
Feminist Data Manifest. [No Manifestno.com] (Website).
Foer, F. (2017). *World without mind. The existential threat of big tech* (1st ed.). Penguin Press – Penguin Random House LLC.
Foucault, M. (1995 [1979]). *Discipline and punish: The birth of the prison* (2nd ed.). Vintage Books – Penguin Random House.
Foucault, M. (2004). Securité, Territoire, Population. Cours au Collège de France, 1977-1978 (EHESS, Gallimard, Seuil).
Graeber, D. (2012). *Debt: The first 5,000 years* (reprint ed.). Melville House.
Ingold, T. (2013). *Making. Anthropology, archaeology, art and architecture*. Routledge.
Kneese, T. (2023). *Death glitch. How techno-solutionism fails us in this life and beyond.* Yale University Press.
Maurer, W. (1999). Forget locke? From proprietor to Risk- Bearer in New-Logics of Finance. *Public Culture, 11*(2).
O'Gieblyn, M. (2021). *God, human, animal, machine: TECHNOLOGY, metaphor, and the search for meaning* (1st ed.). Doubleday – Penguin Random House LLC.
O'Neil, C. (2017). *Weapons of math destruction: How big data increases inequality and threatens democracy* (1st ed.). Broadway Books.
Rabinow, P. (1999). *French DNA. Trouble in purgatory*. The University of Chicago Press.
Shakespeare. (n.d.). *The tragedy of Hamlet, Prince of Denmark.* Folger Shakespeare Library. Accessed April 25, 2024, from http://www.folgerdigitaltexts.org
Susskind, J. (2022). *The digital republic. On freedom and democracy in the 21st century.* Bloomsbury Press.
Warnier, J.-P. (1999). *Construire la Culture Materielle. L'Homme qui Pensait avec ses Doigts.* Presses Universitaires de France.
Weber, M. (2001 [1904]). The Protestant Ethic and the Spirit of Capitalism. Routledge
Zuboff, S. (2019). *The age of surveillance capitalism: The fight for a human future at the new frontier of power* (1st ed.). Public Affairs – Hachete Book Group.

Part III
Where to?

Chapter 7
AI and Diverse Ways of Knowing

In my native Romania, as everywhere, there are stories that precede the modernization of the nineteenth and early twentieth centuries, preserved and transmitted through different means. "Dragobete" (loosely translated as the Loving One) is the celebration equivalent with Valentine's Day. Placed around the full winter moon, around the 24th of February, the legend speaks of Dochia, a very beautiful woman shepherd who slept in the summer outside on the mountain slopes. The mountain loved her so much that it formed a mist around her and impregnated her with a boy that was born in February. The boy was named "Dragobete," and he was so beautiful that both men and women fell in love with him. He was also blessed with a fully lived, beautiful life, reflecting the seasons: a flowery childhood, an expansive summer, a fruitful autumn, and an old age covered by a white diamond shiny coat. Mother earth taught him all the plants and their properties, and he imparted his knowledge to all who loved him. He died becoming himself a plant, Phyllitis Scoloperium, popularly called "The Deer's tongue." The plant is a fern growing in beech forests, known and used for its healing properties: for pulmonary and respiratory tract cleaning, but also as a good diuretic for kidney affections, and for digestive problems. The story has multiple possible layers of interpretation: a reminder of the cyclicity of life following nature's rhythms, a tale of plants' healing properties, the centrality of love, and the importance of living within and not separated from nature herself.

I was raised in an extended family in Romania during the totalitarian period of the 1970s and 1980s, having as primary caretakers my maternal grandparents and living closely to my great grandmother. When I fell sick, there were different home-given healing processes that complemented the "all-use" aspirin (one of the few medications available on the market): suction cups, hot-salted foot baths, chest potato compress, and a variety of plant infusions. Sometimes younger neighbors would come to my grandmother to have suction cups applied. I still have the cups and exercise myself how to apply them to my curious friends. Among the plants, I

A. Balasescu, *Climate Change in the Age of Artificial Intelligence*, Artificial Intelligence, Simulation and Society,
https://doi.org/10.1007/978-3-031-90042-6_7

remember onion or pine-bud infusions or syrup for coughing, chamomile for infections, linden for calming, "rattles" plant for stomach aches, cherry tail infusions for heart problems, and "horsetail" for kidney drainage. The liquid from cabbage pickling was, and still is, used for cleaning and helping the healing of the digestive tract.

My grandmother and great-grandmother knew how to use most of the plants. They did not refuse medical treatment, and they both knew how important it is: my grandmother's sister died of pneumonia only a few years before penicillin became available to the public, a story that was passed on to me. They did not regard the natural path for healing in competition with allopathic medicine. I remember the family doctor herself recommending plant infusions and bodily well-being care, such as hot baths with salt for feet, before any more aggressive, pharma-based treatment. But those were days and places in which pharma was not an industry for pushing painkillers. They were also the days in which medical care was not politicized and society was not polarized like today when the logic of algorithmic communities (Ruckenstein, 2023) pushes toward discourses of purity:[1] either allopathic or "naturist" healing.

During the years before the fall of the Iron Curtain, the Romanian state encouraged the use of plants through a network of naturist pharmacies, aka "Plafar" (an abbreviation for "plant pharmacy") perhaps also in order to supplement the relative absence of industrially produced medicine. Today, luckily, that type of knowledge is not lost. After 35 years, the Plafar boutiques still exist, and people still use plants for healthcare. One can find a wealth of information on various sites that keep that memory alive: packed and presented in a modern marketing manner, with recommendations and know-how sharing.

Throughout our history, knowledge is transmitted, lost, recovered, layered, transformed, and transferred to different regimes of truth. Technologies also migrate and change their use, get adapted to different goals, and ultimately engender new ways of being and doing. They also speak volumes about the societies that use them, only by looking at the logic within which they are used. Take, for example, the camshaft. This is a piece known today for its use in the internal combustion engine mechanism. The camshaft is used to act upon the cams of the engine valves and control their synchronized movements, securing thus the admission and evacuation of air (oxygen, respectively, CO2). It is a translator of movements, laying in the middle of the mechanism (see photo). When it spins, the pistons move up and down, opening and closing the valves. The camshaft is the central piece in the engine, and its rotation is generated by the engine itself through a belt system. It functions in the closed complicated system of the engine.

The camshaft, though, is known since ancient China, and it was quite familiar in the mills during Medieval Ages Europe. The complex technological assemblage and meanings derived depends on what type of system we place it, where we choose to

[1] It is important to emphasize, once more, the danger of politicized discourses on the body, individual and collective, its purity, and its link to nature translated in radical forms of "naturism" and New Agism. They invariably bring in elements or forms of totalitarianism and fascism, by postulating a singular way toward "saving" oneself or the world from the "danger of impurity."

apply the energy force, and of course what type of energy. The water mill used the hydropower to spin the wheel and through the camshaft turn that rotation into translation movement of pistons that would press the material and separate fibers and seeds. The system is complex, open to the environment, dependent on water fluctuations, and in a transformative relationship of "power with" it. The water flows through unaltered, unconsumed. On the contrary, the combustion engine is a closed loop: using the energy obtained through burning carbon-based combustibles it pushes the pistons up and down, and it spins the camshaft that at its turn pushes the valves' cams. It is self-centered and closed, working to master the environment in a "power-over" relationship.

The photo you are seeing (Fig. 7.1) is of a water mill at the "National Museum of the Village 'Dimitrie Gusti'" in Bucharest, Romania. The museum was the brainchild of Dimitrie Gusti, historian and ethnographer, projected and built in the 1930s, after the Romanian Nation-State was established in 1918, and it had a double meaning: it was meant to show the presumed unity of the political territory by creating the category of "Romanian Village" and to muse-ify the "traditions" that where both legitimizing but also "backward" and meant to disappear in face of the celebrated progress and modernization of the country.

Initially the museum was built on the model of the "human zoos" of nineteenth Century Europe—a practice that brought people from the colonies to the European metropole in order to perform their "indigeneity." The first such ideas were talked about in Romanian intellectual nationalist elites starting in the nineteenth century,

Fig. 7.1 Camshaft in a watermill assemblage, the Museum of the Village, Bucharest. Photo by the author

particularly after the Paris International Exhibition from 1867, from which Alexandru Odobescu, a Bucharest intellectual of the era, inspired himself in first proposing an open air museum after admiring the displays of the colonies.

For the European public, this practice created the stereotype of "the other" as both fascinating and threatening, through reifying the binary oppositions that characterized the dominant discourse of the (colonial) times: tradition vs. modernity, magic vs. science, savage vs. civilized, and backward vs. progress. Many times, the people populating these traveling "human zoos" were themselves paid actors from the colonies. In 2022, an exhibition in Bruxelles extensively presented this practice of production of racial and racist discourses, while pointing out also to the voices that opposed from the beginning this abhorrent industry of entertainment (see photos). Although extinct by the 1920s in colonial centers, the newly formed nation-states such as Romania adopted variations of this practice in order to form its own oppositional figure to progress: the "peasant," an ambivalent category that is both the referent to a supposed unifying and all encompassing "Romanian" tradition that legitimizes the territorial political unification and the emblem of the backwardness as dark background for the light of modernization. In the houses of the museum, built in regional styles of the territory but very close to each other, people from those particular regions were brought to live. They were supposed to reflect the harmonious nation living together and thus to demonstrate its legitimacy and its perennial historical character. It was a scaled map of the nation-state territory. It unsurprisingly failed: people who came from different corners of the country and suddenly became neighbors in the miniatured Romania had different ways of doing things and different ways of seeing the world, and unsurprisingly to us but deceivingly for the Bucharest intellectuals and ideologues of the era, they arguably spoke different languages if not radically different dialects of the same language. As the ideological unity of the model demanded the affirmation of uniformity, diversity was ignored and so were its possible effects. After World War II and after being populated for 12 years, it became obvious that the model of Romania built on 4.5 ha of land could not function with living people in it. The simulacra changed its meaning and therefore was about to lose its status. The houses in the museum started to deteriorate because of "the *improper habitation* and practices of raising animals that families brought with them"—the source of the quote is the site of the museum,[2] with my italics Muzeul Satului (n.d.). In other words, people did not know how to live museified. The families originally living in the houses and the war refugees that were subsequently housed there were invited out. The architectural construct stayed, continuing to function as a museum to this day.

There are a few ideas to retain from this story and from the ways of looking at the camshaft. First, models that suppose (human) beings and standardization are doomed to fail if the goal is the perpetuation of the model. Or it implies the transformation of human activities and ways of being to match it, by reducing their variability and diversity, in this case thus revealing the continuous dynamic of

[2] https://muzeul-satului.ro/despre-noi/istoric-muzeul-satului/

subjectivation and opening to the foibles of power exercises. Second, nation-state projects were models that uprooted and homogenized entire populations in Europe, contributing to the loss of languages and ways of being and doing. Third, technology is both neutral and ideological at the same time. The camshaft is not a modern invention, but it is presented as such when it appears in the assemblage of an engine. When museified and exposed in the assemblage of a water mill, it appears as *the opposite* of modern technology: it is traditional, exotic, and backward. It is used precisely to emphasize what is presumably *not* modern, *not* progressive, and *not* scientific but rudimentary and perhaps ingenuous, at most maybe innovative. Most of us do not even make the link between the two mechanisms. The link is not only the camshaft but also the imagination of the humans who built it and who used it. By the same token, the difference is given by the way this use reorganizes the web of relationships involved in the making and functioning of the camshaft: the same technology gains different and radically opposed meanings if used in different regimes of power, energy, and exploitation. The prime material of the making itself suggests this: naturally growing wood versus the extraction-based and energy consuming steel. It also generates completely opposed effects on the environment.

The speed of change and the acceleration of processes in technology, the so-called progress, is only equaled by the speed of change in climate and by that of extinction. What happens generally during those periods, paleontologists argue, is that:

> the usual rules of survival are suspended. Conditions change so drastically or so suddenly (or so drastically *and* so suddenly) that evolutionary history counts for little. Indeed, the very traits that have been most useful for dealing with ordinary threats may turn out, under such extraordinary circumstances, to be fatal. (Kolbert, 2015: 17)

Human activity in the past 300 years is responsible for the current wave of extinction. The Anthropocene introduced a set of variables in the Earth system that rendered obsolete the previous mechanisms of adaptation and change. Yes, there have been five previous instances of extinction. However, the sixth one that began with the Anthropocene is an "unnatural" one, as Kolbert suggests. The latter extend not only to all species but also on ways of being expressed in languages and cultural practices. As climate change pressures species into extinction, a certain type of practices of uniformization demanded by the culture subservient to new technologies exerts pressure on languages and cultural diversity. This is not new. But there is a twist. In the nineteenth century, the extension of railways both in Europe and in the colonies was an instrument of internal and external colonization. In France, the combination of railways with the imposition of a national school program led to the disappearance of tens of local languages and ways of being. Bretone, Occitane, and Basque are the most well known because they are part of active processes of revival in contemporary France. There are countless others. The railways on the American continents accelerated the speed of settler colonialism and literally and effectively provided a fast-moving platform for killing indigenous populations and shooting wild stock, mainly bison. Destruction of diversity happened indiscriminately during the period of the modern empire both in the metropole and in the colonies. However, the twist in the relationship with technology is as follows: while in the nineteenth century, technology was used to serve nation-state projects of population uniformization and various colonial advents (both sides of the same coin), today many societies respond to tech demands of standardization by voluntarily reshaping themselves and their ways of being. Generally they move toward both uniformization of practices (from workflows in organizations to romantic dating, itself subject to workflow logics) and standardization of language that serve the easiness of implementation of technology itself. The reduction of diversity in any system is the first sign of imminent extinction. Technological progress seemed to demand standardization. Diversity reduction became the central ideology, and because of the aura of objectivity, it is not perceived as ideology but as "natural progress." It is well known that from an ecosystemic perspective, the disappearance of diversity is the first sign of the beginning of the dismantling of equilibrium and perhaps gradual end. This is the type of progress that moves us toward extinction. Wade Davis, an anthropologist, advanced the concept of ethnosphere as the totality of the "cultural web of life that envelopes the world," and he affirmed that today its diversity is declining faster than that of the biosphere. In an interview with the National Public Radio (NPR), he says that "the key indicator is language loss. There are at present some 6,000 languages. But of these fully half are not being taught to children. Which means that effectively, unless something changes, these languages are already dead" (Chadwick, n.d.).

The growing interest in indigenous ways of thinking comes with the background of both the movements of indigenous rights and that of the necessary decolonization of thought without which arguably no liberation is possible. Indigenous ways of knowing appear to bring radical answers to the moral, ethical, and relational conundrum that the Industrial vision of nature and the use of technology that it has engendered.

While there is a vast variety of other-than-modern philosophies—from ubuntu in South Africa to First Nation thinkers in Canada or the USA, and the appeal of indigenous worldviews is their holistic, systemic approach that emphasizes relationality over the Enlightenment's excessive rationality. While exploring the instances and propositions of indigenous ways of knowing as they are applied to Climate Change action and AI ethics, we need to consider the strength of their propositions but also of the danger of falling back on the colonial trope of the "noble savage." The argument is that in the process of decolonization of thought, one should be aware that the proposition of indigenous ways of knowing is exactly this: methods of knowing, not a fixed set of knowledge that could be magically transferred and that could solve, say, the ecological crisis. The thinking behind the latter falls rather into the category of the myth of the noble savage that "knows the secrets" and that the "alienated white man" needs to access. Within this type of thinking, again the "indigenous" is constructed as "close to nature," "knowing its secrets," and is ahistorical, since in this view, the set of knowledge is presumed to be fixed, ancient, and immutable. This romanticized vision denies the very power of observation and scientific empirical approach that other-than-modern people have on the surrounding world, as Malinowski suggested. It also would deny the validity of the usage of current tools and technologies to enhance this type of knowledge.

There is also the question of geographical circumscribing and racialization of indigeneity. It is the historical processes and the power relationships informing modernity that uprooted the humans from nature. Indigeneity is a category linked to the colonial enterprise that spread the modern ideology by violence and force outside Europe. However, within its geographical boundaries and at their margins, such as in Romania, local forms of knowledge have been suppressed by similar means, both symbolically and physically. "The peasant" in the national museums dedicated to it, and "the indigenous" are both categories that mean the same in the colonial modernizing discourses. It is important not to fall into the trap of reproducing this system of thinking by simply opposing it. This is why using the encompassing terms such as Indigenous and Local Knowledge (ILK) or other-than-modern ways of knowing, instead of the geographically circumscribing ones such as "Western" and "non-Western," may also point to a way to express the extinction and what precisely is threatened: the knowledge of the land and of the intimate link we have with it. In their article *Scientists' Warning to Humanity on Threats to Indigenous and Local Knowledge Systems* (Fernandez-Llamazares et al., 2021), the authors highlight the pitfalls of a romantic infused view that informs the framing of the issue solely in terms of loss and necessity of "conservation":

> In particular, focusing on losses does not do justice to the immense and powerful continuity that is a hallmark of the cultures of Indigenous Peoples and local communities [...]. Similarly, the dynamic nature of ILK systems can be overlooked in culture "preservation" plans, which often purportedly aim to fix ILK in place and time [...]. Furthermore, the narrative framing ILK losses is also often problematic, as many communities affirm that their knowledge systems were not simply "lost," but rather violently wrested from them and destroyed with intent. (Fernandez-Llamazares et al., 2021: 147)

Any type of human knowledge has both a history of accumulation and a historical change process. The importance of other-than-modern ways of knowing in this moment of planetary history is also revealed by its methods and by the proposition of holistic relational understanding that can be applied to any human endeavor (and it is not limited to "knowing nature" differently). Approached from this perspective, not only the climate change challenge but also our understanding of AI, may gain whole new dimensions. Modernity discourses discarded the ILK systems and deemed them unscientific. This is an ideological stance, as previously discussed. There is nothing unscientific in the way indigenous and premodern populations approached the world and acted in relation to them. The fact that it is not formalized in modern mathematical language but preserved in rituals and oral transmissions does not make this form of observation less scientific. Again, a century after Malinowski made this case, the world is still caught up in this binary opposition between positivism considered objective science and the presumed magical state of any other form of knowing. The positivism that extended to all sciences in the nineteenth century and the dataism that eats away at the balance given by the qualitative approaches to our reality helped perpetuate this discourse of invalidation of other-than-modern ways of knowing. The difficulty of understanding relationality and the projection of a mathematical system on the dynamic character of the world give birth today to practices such as conservationism that are criticized by indigenous scientists. Jessica Hernandez formulates very clearly this point when talking about invasive species:

> The Western sciences teach us that invasive plants are pests, unwanted, or do not belong in this landscape. However, to us, invasive plants are displaced like many of us. They were forced from their native lands and like many of us had to adapt to a new environment. Like banana trees, we are forced to uproot ourselves from our native lands and have to adapt to new environments. Some invasive species have thrived so well that many folks think they are native species because of the abundance of them in many landscapes. While not forgetting the negative impacts they do have on some environments, some of our Indigenous communities have learned how to live with them. (Hernandez, 2022: 37)

In her book, the Oaxacan biologist provides a forceful critique of conservationism, tracing its history to the colonial period and showing the parallel between environment conservationism and the attitude toward indigenous people who engendered the ideas of preservation and also reservation. This view imposes a static character to reality, and this of course corresponds to the deeper belief of a mechanistic universe formed by elements that function in aggregate but that are immutable in themselves. Ultimately, despite the appearance, this is a rather religious type of view, in which it is postulated an ontological quality for each element that forms the universe, a fixed one (possible to quantify this), and eventually appearing in a pure state. This lies also at the basis of the myth of exact science (singular) that needs the immutability of elements that are to be measured and brought into the mathematical language. The extreme view of this is the attempt to transform everything into data. The purity though is a characteristic more akin to monotheistic religions and also of the Platonic view of our world as the mere reflection of the perfect world of ideas.

A telling deconstruction and critique of this view is offered by Mary Douglas in "Purity and Danger," in which the author explains how the categories of pure and impure intertwine:

> [I]n general the underlying principle of cleanness in animals is that they shall conform fully to their class. Those species are unclean which are imperfect members of their class, or whose class itself confounds the general scheme of the world. (Mary Douglas, 2012 [1966]: 55)

Douglas goes on to argue that impurity represents danger in this worldview, as it menaces the vision of the rigid order of the world, that is of the perfection of a static creation that would make the universe function as a mechanistic assemblage of perfect categories. Those who are impure in a system of categories are deemed to be "dangerous." This passage extracted here refers to Leviticus and food impurity. In what I prefer to call the identitarian (dis-)order of things, we can replace in this text "animals" with humans and "species" with ethnic, religious, racial, or any other type of prevalent category that appears in discourses that rigidly categorize "other" groups. Thus, if we reread the passage in terms of conservationism, we can see the trouble with it, and we can also recognize the bigger trouble that Hernandez is making us aware of; the incapacity to politically accept the dynamic character of, well, everything, identity included:

> In general the underlying principle of cleanness [in humans] is that they shall conform fully to their class. Those [groups] are unclean, which are imperfect members of their class, or whose class itself confounds the general scheme of the world.

The practices this view engenders is what Bruno Latour calls a bicameral vision of the world, with two types of separations. First, there is a separation between humans (ethnocentrically considered, that is, the colonial Europeans) and nature, with the indigenous people viewed as a link between the two, a vision that transpires today in some discourses about indigeneity.

Second is the separation between science (singular) and all other forms of knowledge, in which "the scientist" (singular) is the person who has access to the "real world" of ideas aka objective reality and who can translate it to our world that is only an imperfect projection of what ought to be. Remember Leviathan?

They are combined in the politics of nature that places Europeans as always already separated from nature or as Latour says "they have never been living in nature" and the indigenous as *only* living harmoniously with nature. These are "two symmetrical exoticisms" at work that keep the category of "nature" itself in place, but by doing this, they perpetuate the separation from it and the separation between those who "know nature through Science" (singular) and those who "are (in) nature."

> We should be able to get along entirely without the *two symmetrical exoticisms*: the one that makes Westerners believe that they are detached from nature because they have forgotten the lessons of other cultures and live in a world of pure, efficient, profitable, and objective things [nature included as "things", my note]; and the one that made other cultures believe that they have lived too long in the fusion between the natural order and the social, and that they needed finally, in order to accede to modernity, to take into account the nature of things "as they are." (Latour, 2004: 46)

In other words, what keeps us separated from nature and among ourselves is a belief anchored in rather religious principles presented as "science" (singular): a perfect, ideal, objective world separated from a nature that hides its own truth from everyone's eyes except those of the initiated one, because the truth about the world resides in the sacred sphere of ideas. Modernity is thus discursively presented as the perfect remaking of that ideal sacred world for us to live in and enjoy its fruits, as long as we accept the separation from nature, and math as its perfect language. Sciences, plural, are the domain of trial and error, of empirical observations, of both quantitative and qualitative assessments, and of engaging meaningfully with the physical world through making and are characteristic to *all humans.* But, in modernity's conception, they are both regimented into separate disciplinary fields that seek legitimacy through placing themselves in relationship with "hard sciences" and *denied to those who do not use the method of separation and control through numbers.*

In this context, one can even better understand the appeal of the promise of artificial intelligence discourses: through data extraction, everything can be knowable in terms of science, not of sciences. The world "as it is" will reveal itself through the algorithms. In this vision, the data scientists are those who discern the ultimate truth about reality, anything less than this in terms of knowledge being deemed inefficient, impure, unprofitable, and of course subjective. The problem with other-than-modern types of knowledge is not their validity but their lack of legitimacy within this regime of truth.

In their article, Fernandez-Llamazares et al. use three case points to illustrate the deep knowledge about the world embedded in cultural practices of management and governance of resource use and exploitation in ILK: the NW Pacific Coastal Indigenous populations in relationship to the herring, the ethnobotanical knowledge of Tsimane' people in currently Bolivian Amazonian basin, and the Maori knowledge for conservation of Burrowing Seabirds in New Zealand. They illustrate three different instances of relationship within the world: that of more populations with one species, the herring, that of one population to many species in Amazon, and that of one population with one species in the case of Maori. They all point to similar conclusions: indigenous and local sets of knowledge embedded in cultural practices are nothing less than scientific ways of seeing the world resulting in rules of governing its interrelated gears. They are the result of continuous observations and accumulations of a vast body of knowledge, with rules of conduct formalized in rituals and cultural practices. They embed forms of protocols that are proven to be a way to secure resources and also to adapt to changes. They are not unscientific by any means. They are though not formalized in mathematical standardization. And this may in fact be an advantage. The downside of quantification in complex systems is, as argued before, the severing of the interlinks between the elements of a system. What remains visible is only what is prone to quantification. Principles of unity and reciprocity in the system are sacrificed in order to make its constitutive elements measurable. And this leads many times to the loss of understanding of the webs of relationships and their potential of explainability of the phenomenon as a whole. If we do not make the quantification the scope itself, but one of the multiple means of understanding and being in the world, chances are that we will be looking differently at our tools and particularly at AI. It is worth remembering here the words attributed to Niels Bohr, the quantum physicist: "There is no quantum world. There is only an abstract quantum physical description. It is wrong to think that the task of physicists is to find out how Nature is. Physics concerns what we can say about Nature" (Petersen, 1963: 12). In his article about the philosophy of Bohr, Aage Petersen (ibid.) reminds the reader that the mode of understanding of the complexity of the world depends on the tools used for this understanding and that quantum physics, while using mathematics, does not make the confusion between the exactitude of the system and the accuracy of the description. On the contrary, it reifies the dependency of description to the method and affirms that "[q]uantum physics showed that even the causal mode of description is only one special way of tying together the natural phenomena, and thus lawfulness in nature is not equivalent with mechanical causality" (Petersen, ibid. 11–12).

Speaking of tools, Andrew Zolli from Planet.com exemplified to me how technology may help indeed in raising planetary consciousness or at least illustrate how indigenous stewardship is needed in zones of crisis. During one of our interviews, he showed me images of the Amazonian jungle taken by the minisatellite cameras. In the sea of green, there were some wondrous lines that I thought were rivers. He said: "These are roads, illegal roads. Look what happens!" The next set of photos of the same coordinates, taken 3 months after, showed the complete erasure of the forest on big surfaces around the initial road lines. However, a little further and yet in the vicinity, there were vast surfaces that remained untouched. “These are lands given to the local indigenous people to care for. As you can see, from 1984 to 2018, while everything else is deforestated, these lands stayed untouched” (Fig. 7.2a, b).

The conversation was taking place in the final years of Bolsonaro rule in Brazil, during which the Amazonian forest became open to exploitation, legal and illegal. It would be naive to think that the defense of the forest by indigenous population

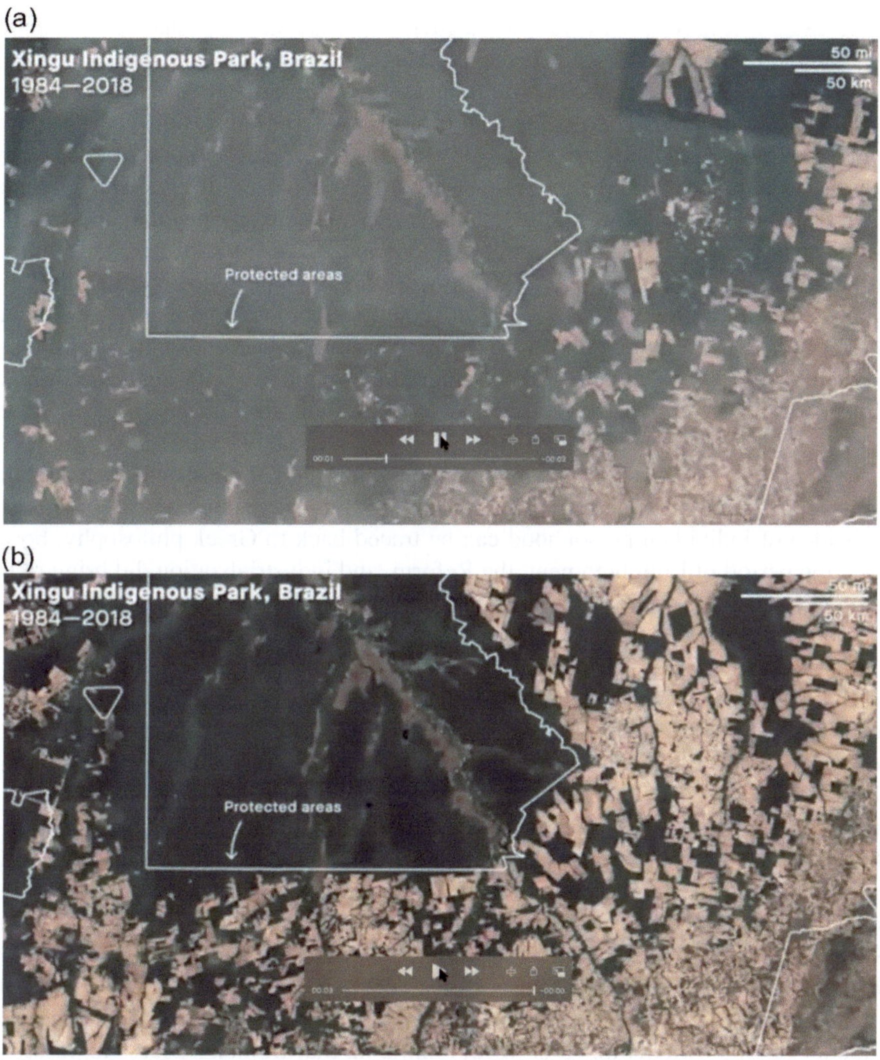

Fig. 7.2 Amazon deforestation sequences between 1984 (**a**) and 2018 (**b**) showing the positive impact of indigenous land stewardship. Courtesy of Andrew Zolli

happens without conflict and violence. Nation-states independence from colonial powers in the nineteenth and twentieth centuries did not mean necessary decolonization but many times the shifting of power of extraction and exploitation to new actors, the nation-states, within the same colonial logic. Decolonization must be within the ways of being, doing, and understanding of ourselves, the world and our place in it.

At this point, we can return to the story of the camshaft. Up to now, we have looked at AI in the way we regard the camshaft in the assemblage of an internal combustion engine, functioning with the political economy of late modernity and the mechanistic view of the universe, individualist personhood, and computational mind. But, as it is with the camshaft, artificial intelligence could and must be integrated in a different set of practices and worldviews. In most of the indigenous and local knowledge systems, the central point is not necessarily the individual person but the relational one. That is, personhood is considered the result of the web of relationships that characterizes the world and not the generating point of it. In other words, the human *is not* "the measure of everything and every thing." The relations are a fact also reflected, for example, in the forms of salute among many indigenous people living in today's Canada: "all my relations." Those relations are not limited to the interhuman ones but extend to the world as a whole. They are the fabric of subjectivation; the individual as "person" emerged from postulating the severance of some of these relations and from the emphasis of others, through disciplining Foucault observed: "The individual is rather a certain mode of carving the multitude [...] rather than the prime-matter out of which multidute is built [as Hobbes would have it, my note]" (Foucault, 2004:14).

In the previous chapters, I touched on the matter of what constitutes personhood in the (post-)Enlightenment period and how individual personhood is produced. This idea of individual personhood can be traced back to Greek philosophy; however, the period of Enlightenment, the Reform, and industrialization did bring radically new meanings to it: ownership and possessive individualism. We operate with variations of these in the modern paradigm. The ideas of author, authorship, and authority are strictly linked, and they have to do with the emerging ideas of property in the philosophy of Enlightenment (eighteenth century)—particularly British philosophy. Needless to say this stayed at the basis of building the capitalist and colonial system that continues to influence us today. Usually indigenous and traditional forms of knowledge do not emphasize authorship, for example, indigenous or traditional art is not authored and the practice of authoring the indigenous art was introduced by the colonial presence. Also it needs to be reminded that authority is a form of power acquired by appropriation—taking possession, on which the colonial advent is based. By the way, in Hong Kong, there is a street called "Possession Street." It is the first street built by the British and it marks the taking possession of the Island. It was the first way of access from the harbor to the core of the Island. In one of his lectures on colonialism, Achile Mbembe remarks that "colonialism is first a matter of land ownership." It is worth reminding that women's access to possessing property (and also recognition as authors) was very, very late, as was their access to education. Many women writers needed to use men's pen names in order to have access to being published. Some scholars argue that Shakespeare himself, the epitome of "the author," may have been a pseudonym for Emilia Bassano—a Jewish woman born in London from a Venitian immigrant family (Winkler, 2019).[3]

Following this train of thought, it is also interesting to see how in current language ideas and things are attributed authorship or "paternity" but not "maternity"—language again showing assumptions deeply ingrained in our ways of being and doing. These ways of thinking and doing were fairly new at the time, and they start configuring the framework of power and subjectivation that created the idea of the individual, that is, the individual as a separate entity from his (not her, because at the time the standard was "the man") family and community, endowed with individual will and agency, and necessarily holding property over his goods or ideas, be them land, house, and what they contain (including women and children), rights (the rights are to be possessed), or the very grass that one's horse is grazing (as David Hume put it). When Michel Foucault is talking about the productive character of power, very often he refers to this genealogy of power, and how its most interesting product is the individual itself, produced through practices of individuation and individualization, present in schools, churches, prisons, hospitals, and other institutions that mediated the relationship between people, their bodies, and their governing bodies. It is one thing to stay in a group and in a circle when being in the process of learning, and it is completely a different thing to be individualized on single chairs, assessed on your individual performance and constantly prompted to perform as an individual. The same goes for hospital beds. We take these things for granted today. They were not 200 years ago. One of the most important new instruments of the power of individualization was rendering the individual "visible"—in the literal sense of the term. The whole overhaul of philosophy and physics during the Enlightenment was based on the idea that visible means truth (this is obvious even in the name the thinkers of that time attributed to themselves—"Philosophers of the Enlightenment," meaning bringing to light). The power mechanisms embodied in practices of that time, and then on, are intimately linked with the equal sign between visibility and truth.

This is the political economy of the personhood, centered on possessive individualism (MacPherson, 1964), that forms the engine in which "the camshaft" of AI is the central piece. Silicon Valley's culture, out of which the dominant view on AI emerges, functions on a particular and extreme declination and expression of this type of personhood. The pistons that train AI into movement are efficiency, objectification, individualization, profit, competition, etc. These are ways of organizing the world that engender the type of behaviors that accelerate climate change. Climate change is not necessarily a function of technology itself but of the behaviors and worldviews in which this technology is integrated and that in turn it generates or solidifies. In their article *World scientists' warning: The behavioural crisis driving*

[3] Some scholars argue that Shakespeare himself, the epitome of "the author," may have been a pseudonym for Emilia Bassano—a Jewish woman born in London from a Venitian immigrant family (Winkler, 2019) https://www.theatlantic.com/magazine/archive/2019/06/who-is-shakespeare-emilia-bassano/588076/

ecological overshoot, Merz et al. argue[4] that there is a common set of behaviors entrenched in consumer capitalist cultures that drives the planet to the limits of its survival. These are the fixation on economic growth, marketing as a driver of consumption, and pronatalism politics and policies. The authors make the case that the first one drives the two subsequent behaviors and that the solution to climate change is indeed behavioral. In a dystopian reality, this may even look like the USSR proposed by Roy Bendor and myself. However, what if one of the keys is really looking at what person and personhood means and how the alienation of the self from the world that Hannah Arendt argued characterizes modernity and that drives the way we understand and relate to technology, is just one among many possibilities? What if instead of pushing these pistons to spin the AI economy, we start by spinning the camshaft by other means?

This is the type of proposition that Sabelo Mhlambi, a computer scientist, researcher and Fellow at Harvard's Berkman Klein Center for Internet & Society and a Technology and Human Rights Fellow Carr Center for Human Rights Policy, is making through his work. In his paper *From Rationality to Relationality. Ubuntu as an Ethical & Human Rights Framework for Artificial Intelligence Governance*,[5] Mhlambi details how the principles of Ubuntu philosophy shed a new light on the possible use of technology. At their center, there is the relational definition of personhood that decenters the importance of the individual and places the person in the web of communitarian, social, material, and historical relations that converge into a person. This would engender a completely different ethical agenda for the governance of AI and digitization, in which the person is dignified through the giving of importance to the web of relations and values that make that person whole. In other words, it is not "the subject" but the ways of subjectivation that should be at the center of ethics—in AI and elsewhere:

> Tech companies seeking to create technology to achieve ethical outcomes should consider the following ubuntu principles: solidarity, reconciliation, equality, equity, and community. Through solidarity, technology that creates social cohesion should be pursued. Reconciliation should be a practice that places disenfranchised communities in positions of power within the company and in society. The principle of equity mandates that technology companies reduce inequality through their product offerings. The principle of equality is rooted in human dignity, and mandates that human rights protections should shape technology. The principle of community encourages tech companies to give greater control of the product and its use to the community. (Mhlambi, 2020: 25)

To achieve this, Mhlambi proposes a set of very clear steps to be applied in product design and usage, from community ownership of data that confers control over it and the possibility of revoking the rights of use for IT companies to adapting

[4] Merz, Joseph et al. "World scientists' warning: The behavioural crisis driving ecological overshoot" in Science Progress 2023, Vol. 106(3) 1–22.

[5] Sabelo Mhlambi (2020) "From Rationality to Relationality. Ubuntu as an Ethical & Human Rights Framework for Artificial Intelligence Governance." in Carr Center for Human Rights Policy, 2020 (9) pp. 1–27.

algorithms to social and cultural normativity and subjecting their recommendation to "ideals of human dignity and social cohesion."

The principles of this type of approach are to be found in other-than-modern ways of thinking or in ILK systems. The removal of rationality as centerpiece of the social imagination and its replacing it with relationality at the center of ethics is fundamental because it proposes a way of understanding personhood that was actively forgotten in the colonial and industrialization periods: the relational personhood, an understanding from which everything else derives. There is an urgent need to consider them also at the core of how technology is designed and used. The "Abundant Intelligences Project"[6] is an international indigenous-led project on AI that unites researchers and fellows from Canada, the USA, and New Zealand and aims at Indigenous People's full participation in technology shaping and design (New Frontiers in Research Fund, 2023). The project is co-led by Jason Edward Lewis (Concordia University, Canada). Lewis identifies clearly that AI does not have an ethics problem but an epistemological one:

> [It] creates huge gaps between the diverse forms of human and nonhuman intelligence and what the AI industrial-academic complex is trying to replicate. Collapsing intelligence down to a rational, goal-seeking, self-serving agent—in all of our cultures, that kind of person would be seen as selfish, foolish, not a good community member, and not intelligent. (Lewis in Social Science and Humanities Research Council of Canada (2023), accessed February 2024)

Hēmi Whaanga (School of Māori Knowledge, at Massey University in New Zealand), colead investigator, explains that the project aims "to mobilize AI technology to explore the centrality of hapū (kinship group) and connection in creative, language, cultural and well-being contexts" (idem.), where well-being is understood in conformity with the principles mentioned. Within this framework of understanding, well-being thus is itself relational and derived from the placing of personhood in healthy relations with her context, kinship, community, history, ancestry, and future (Lewis, 2020).

Well-being is thus not denying individual death by means of salvation into the machine Kurzweil style or by minute measurement and management of biology through technology, as Bryan Johnson suggests in his quest for immortality. Well-being is understanding the individual within, and thus replacing it into, the greater assemblages of the cyclical aspects of its own biology, enhanced as it may be, but still individually finite. Securing our collective survival ultimately means accepting our individual finitude.

This is what it would mean to spin by different means the camshaft of AI: not push it into spinning with principles of competition and centrality of a presumed autonomous rational individual, in a closed, self-centered system. Instead, we need to allow it to turn with the greater cycles of life, time, and relationality that would naturally allow individuals to fulfill their potential within the world and not separated and alienated from it. It would also move from static identification of a person

[6] https://www.indigenous-ai.net/position-paper/

by surface characteristics such as gender, skin, residency, transversal ethnicity, or momentary social status—all categories with which current classificatory technological engines operate and that they reinforce together with all their biases and exclusionary practices—to deep identification by personal histories, deep ancestry, placement within the community, and care for the future with profound political implications. I will return to this in the last chapter of the book.

At this point in the book, we can read it as presenting two opposing sets of principles: the individual centered one, based on disembodied rationality, or the philosophy of (Western) modernity, and the other-than-modern, non-Western and/or indigenous set of philosophies that operate with a different concept of personhood in which the subject/object opposition does not make much sense. But we can also read it as a way to build on the commonalities of life and what it means to be human, if we do away with the concentration on individual humanness and concentrate on values, feelings, relationality, and subjectivation.[7] Some Western philosophies contain forceful critiques of Cartesianism and affirmations of centralities of relationship, as I tried to show during the previous chapters: feminism, theories of subjectivation, actor network theories, and most of anthropological thought in fact attempt to marginalize and particularize the presumed universality of modern Western philosophy and to problematize this geographically assigned category. On one hand, local Western knowledge may preserve, under the flat surface of shiny modernity, practices that are based on holistic relational understanding and placing oneself within the interlinked web of our lives. On the other hand, Indigenous and Local Knowledge Systems do not deny the individual, as some would be tempted to say, particularly those who find themselves influenced by capitalist neoliberal concepts of human beings and libertarian philosophies. Instead ILK systems *replace the individual within the web of relationship* that creates its subjectivity, that is, in the fold of subjectivation. If we are able to profoundly relate to this, another binary that contains strong judgments and hierarchies of societies would become irrelevant: that of individualism vs. collectivism.

In *Principles of Tsawalk. An Indigenous Approach to Global Crisis*, hereditary Chief Umeek/ E. Richard Atleo (2012) puts forth a series of principles of governance from a *Nuu-chah-nulth*[8] perspective:

> One of these principles is recognition. It is to be hoped that, in time, the relationship between indigenous and dominant societies will be marked by experience of and knowledge about each other, which at least would offer the opportunity to develop an ability to recognize the value in each other. (Umeek, 2012: 91)

To achieve this, another fundamental principle needs to be observed: the principle of consent that would allow "by community agreement, freedom of movement,

---

[7] Jimena Garcia Chalci (2023) developed a theory of leadership based on the centrality of relationality, both transversal and transgenerational, in her MA thesis în Global Leadership at the Royal Roads University – 2023.

[8] The *Nuu-chah-nulth* are a group of closely related communities on Western Vancouver Island, British Columbia.

freedom of speech, freedom of thought, freedom to be unique, freedom to grow, and all those freedoms-of-being that do not violate the freedom of others to be" (ibid. 97). Affirming one particular form of life without denying others is indeed possible but only if fundamentally we recognize that there is indeed no "otherness." And this brings Umeek, and us, to the principle of respect and continuity:

> *Iis?ak*, sacred respect, is consistent with the Nuu-chah-nulth assumption that all life forms have a purpose. The existence of life forms is given value by K waa?uuc, Owner of All. Value exists by association with *K waa?uuc* and, thus, all life forms have value and all are to be allowed to continue to live sustainably because of this value. Consequently, continuity is a fundamental tenet of *Nuu-chah-nulth* constitutionalism, which was set into practice through the development of protocols between peoples, nations, animals, and other life forms. (ibid. 117)

Continuity is also the principle that forms the basis of understanding through the lenses of relationality, the principle that von Humboldt invoked when he affirmed the important matters to the soul that "escape measurement" and forcefully dismissed slavery and colonialism, the principle evoked by indigenous women and Western feminists in defense of body and territory, and also the principle that informs practices that are deemed "backward" within current modernity's framing, because modernity, for all its achievements, and despite its rhetoric, does not follow the principle of consent when it marginalizes all other practices and ways of being deemed nonmodern. At the same time, modernity contains its own contradiction and makes possible thinking away from it.

As I was recounting at the beginning of this chapter, my grandmother would take care of my sickness using both modern and other-than-modern practices, with good results. I am alive and healthy. Sometimes in the evening, when I was feeling really dizzy from my sickness, she would slip in a chant against evil eye. It involved burning a couple of matches above a glass of water while singing a song. She would drop the still burning matches into the glass and watch if they float or sink. Based on this, a decision was made regarding my state of being and further treatment which usually included doctor visits the next day. I would then drink the water. It invariably made me feel better (or at least this is what I remember now). I am not saying here that the water gained miraculous healing properties. I am relating how practices may trigger somatic effects that may lead to the amelioration of the state, that is, placebo. This year I heard a story of a more complex and wonderful practice that reminded me of this. I recount it here with the permission of the person who shared it.

Gayathri Gopan is from Kerala, India, and she currently studies in Canada. During a session of presentation of self through a significant object in the core course I am teaching at Royal Roads University, she brought in a *Vilakku* and a *Kinddi*. A Vilakku is a burning vase that uses oil and a cotton thread, while a Kinddi is a vase of water. Ghayathri recounted how when she was a child in her village, Malayankeezh, her grandmother Bhagavathi Amma would light the cotton of the Vilakku every evening, and invite her to sit quietly and think about the day that had passed. The grandmother would pull the cotton thread for burning from a corner of the sari she would wear everyday. Near the Vilakku, the Kinddi with water would be placed. After the worries of the day would be quieted and burned, Bhagavathi Amma

Fig. 7.3 Bhagavathi Amma, Gayathri's grandmother—photo courtesy of Gayathri Gopan

Fig. 7.4 Gayathri's mother and daughter continuing the practice—photo courtesy of Gayathri Gopan

would take the Kinddi and use it to water the plant that grew at the entry of the house. The plant is known under the name Thulsi, and its leaves and branches are used for medicinal purposes to heal the family members with infusions, whenever necessary. Every house in the village has a Thulsi at the door, watered with the Kinddi that sits by the burning flame of a Vilakku. The practice continues in the family, even far away from home. Gayathri's mother, a grandmother for Gayathri's daughter, follows the same practice with her granddaughter (Figs. 7.3 and 7.4). In the light of the principle of continuity, the story is self-explanatory.

Placing ourselves in the midst of interrelated webs of being and doing may give new dimensions to our actions and new purposes to our technologies. The following chapter illustrates how a reversal of perspective on technology may rearrange our approach to sustainability and climate change, possibly generating cascading effects on the assignment of meaning for categories such as value, leadership, and finance. It also may point to possibilities of governance that take into consideration all forms of life and being based on the principle of continuity.

References

Chadwick, A. (n.d.). *An interview with anthropologist Wade Davis* [NPR broadcast]. Accessed April 12, 2024, from https://legacy.npr.org/programs/re/archivesdate/2003/may/mali/davisinterview.html

Douglas, M. (2012). *Purity and danger: An analysis of concepts of pollution and taboo*. Routledge Classics – Taylor & Francis.

Fernández-Llamazares, Á., Díaz, G., Garvin, T., Reyes-García, V., Bridgewater, P., Castro, A. J., et al. (2021). Scientists' warning to humanity on threats to indigenous and local knowledge systems. *Journal of Ethnobiology, 41*(2), 144–169. https://doi.org/10.2993/0278-0771-41.2.144

Foucault, M. (2004). *Securité, Territoire, Population. Cours au Collège de France*, 1977–1978. (EHESS, Gallimard, Seuil).

Garcia, J. (2023). *Relational wisdom practices for planetary health leaders* [Master's thesis, Royal Roads University].

Hernandez, J. (2022). *Fresh banana leaves: Healing indigenous landscapes through indigenous science* (1st ed.). North Atlantic Books.

Kolbert, E. (2015) *The sixth extinction: An unnatural history*. Henry Holt and Company.

Latour, B. (2004). *Politics of nature: How to bring the sciences into democracy* (1st ed.). Harvard University Press.

Lewis, J. E. (2020, September 10). *Indigenous protocol and artificial intelligence*. The Initiative for Indigenous Futures and the Canadian Institute for Advanced Research (CIFAR).

Macpherson, C. B. (1964). *The political theory of possessive individualism: Hobbes to Locke*. University of Chicago Press.

Merz, J. J., Barnard, P., Rees, W. E., & Smith, D. (2023). World scientists' warning: The behavioural crisis driving ecological overshoot. *Science Progress, 106*(3), 1–22.

Mhlambi, S. (2020). From rationality to relationality. Ubuntu as an ethical & human rights framework for artificial intelligence governance. *Carr Center for Human Rights Policy, 2020*(9), 1–27.

Muzeul Satului. (n.d.). *Website of the Museum of the Romanian Village*. Accessed February 1, 2024, from https://muzeul-satului.ro/despre-noi/istoric-muzeul-satului/

New Frontiers in Research Fund. (2023, March 30). *Indigenous-led AI: How indigenous knowledge systems could push AI to be more inclusive*. Government of Canada (canada.ca)

Petersen, A. (1963). The philosophy of Niels Bohr. *Bulletin of the Atomic Scientists, 19*(8), 8–14.

Ruckenstein, M. (2023). *The Feel of Algorithms*. University of California Press.

Social Sciences and Humanities Research Council of Canada. (2023). *Indigenous-led AI: How indigenous knowledge systems could push AI to be more inclusive*. [Government of Canada website].

Umeek, & Atleo, E. R. (2012). *Principles of Tsawalk: An indigenous approach to global crisis*. UBC Press.

Winkler, E. (2019, June). Was Shakespeare a woman? *The Atlantic*. Accessed April 25, 2020, from https://www.theatlantic.com/magazine/archive/2019/06/who-is-shakespeare-emilia-bassano/588076

Chapter 8
The Nonhuman Stakeholders

To act is easy. To think is hard.
To act according to one's thinking
Is troublesome.
J.W. von Goethe.

I spent most of my undergraduate years in a small apartment under the roof of one of the French-style architecture buildings in Bucharest, Romania. The roof windows would allow me to watch the changing sky, to enjoy the shade of sparse clouds over the blue summer skies or the raindrops whose sound would accompany Miles Davis' trumpet, to count the stars in appearing in the relatively small rectangle of the night, or to spy on the unsuspecting people in the neighboring building. By far the most spectacular view I had was the daily crows' migration into and out of the nearby Cismigiu Garden—the central park of Bucharest. Every morning, every night, thousands of crows would fly over the roofs of Bucharest, going to and from between their nests in the park and one of the city trash dumping centers, named "trash holes," situated at the outskirts. Their calls would wake me up with the first rays of Sun in the morning, when they were leaving, and would announce to me the end of the daylight in the evening. The view was spectacular, their flying bodies covering the sky as far as one could see, their overwhelming voices overriding the music or voices of conversation in my house.

In the evenings, they would spend a few hours in the nests on the tops of the trees, chatting with each other about the day, I would imagine, feeding the chicks if there were any, and arranging the nests for the night. In a vision perhaps guilty of anthropomorphization, they were following the same life cycles of commuting humans: wake up, go to work, come back, chat, and sleep. They would do this collectively.

There was something else that they would do and that pushed my reflection about the cyclicity of processes and their invisible entanglements: from the heights of their nest, the crows would defecate abundantly, covering the ground and the park

A. Balasescu, *Climate Change in the Age of Artificial Intelligence*, Artificial Intelligence, Simulation and Society,
https://doi.org/10.1007/978-3-031-90042-6_8

alley with locally produced guano. They were, in fact, processing part of the city's waste and bringing it back into the urban fabric. The circle was closed. As if the birds did not agree with the waste management of the city and reminded the human inhabitants that after all it was their refuse, they should not discard it at the outskirts. They were pleading for a better solution. At some point, a solution came, but it was not about the waste. In the mid-2000s, the newly elected mayor, soon to become the president of Romania, decided that "enough was enough." Small electric cars with loudspeakers began passing every night, playing crow cries equivalent to alarm sounds. The birds were agitated and could not sleep, flying over the trees and crowing in their turn. The whole brouhaha bothered crows and citizens alike. After a few months of futile experimentation, the initiative was abandoned. To this day, the humans continue to deposit their trash outside the city, and the crows patiently carry it back, processing it on the way.

The model of the city that administrators still present to the citizens is that of a closed and heavily managed system. Even if and when sustainable, and rewilded cities are in focus, that wilderness is tamed; it is *nature* in the epistemological sense of Enlightenment, with the centrality of humans in the model. And so it is for all models generated within this episteme, including those used for AI: a reality operable within the sphere of mathematics, manageable and rather mechanistic. For all its flatness, the current modeled reality perpetuates the hierarchy that places humans at the top. Therefore it also generates the fear of "replacement" by machines, in this view a better, more apt human. If we want to have a data-generated model as close as possible to reality, we should not privilege some data generating nodes over others—in other words, humans over other beings.

To achieve this, we need to move as close as possible to a mindset that conceives of humans as an equal part of the environment—the move from egocentric to ecocentric thinking. A true ecocentric view, in which humans cease to be the gravitational center around which everything else revolves, can be achieved only through the practice of placing human and nonhuman stakeholders on an equal footing. This chapter is a return to "nature" in the guise of considering the nonhuman stakeholders of general digitization.

The digitization of everything that transforms objects, beings, and phenomena in data generating nodes signals our belief that nature is just chaos and we can do better for nature than nature does for herself. But this also begs some fundamental questions: who are the stakeholders in this process? If climate change action is sorely needed, how is it to be conceptualized in data-based systems and models? Who are the stakeholders of the action, and how might they be (mis-)understood? How do different approaches to various solutions to climate change take into consideration (or otherwise) nonhuman stakeholders? For example, could business models and approaches change when interested nonhumans are anthropomorphized to a greater or lesser extent; when we think of rivers or forests as equally important as stakeholders? It is envisageable, and the chapter will illustrate with a concrete example, to have bees and sheep in the center of a business stakeholder map. By necessity, this changes the mapping of the business itself, with a lasting impact on how we relate to business models in the first place. The same effect can be achieved through our understanding of digitization. Based on the concept that agency is not exclusively human, I will emphasize the importance of nonhuman stakeholders in creating and applying models of action for climate change and beyond, particularly when using data-based systems of extraction and decision-making as a basis for this action. But is this currently acceptable?

The Case of Pollinators

In 2007, I met Florin Furdui, a young (at the time) office building developer. He was running the regional office of a multinational real estate development company that built a sizable part of Bucharest's business park. Florin's vision can be credited with the first office buildings in Bucharest that use sustainable energy sources, such as geothermal pumps generating around 60% of the offices' needs. Although coming from different worlds, me being at the time a young person freshly out of graduate school, we forged a friendship based on commonalities of concerns regarding life on Earth and what humanity means. The friendship stood the test of time and that of geographical distances that separated us due to my nomadism.

In 2020, after the pandemic set in, Florin told me that the current office building he was engaged with at that time was the last one and that he was learning at great speed the energetic transition ropes, because he was intending to develop land for photovoltaic plants. He became very successful at it; up to the moment of this writing, his company already developed and partially built solar plants on more than 1000 Ha of land. I was living in Frankfurt at the time, and we kept our conversations alive on messenger services. Around the same moment in the story, he sent to me a petition by Avaaz.org addressed to the EU Commission concerning the use of pesticides and other measures that threatened the livelihoods of bees and wild pollinators. The bees were used as charismatic animals to mobilize citizens for the larger cause of the pollinating insects without which life, as we know it, would not be possible. I wholeheartedly signed the petition and also could not help but write back to Florin something along these lines: "I will sign all the online petitions possible, but you are developing a lot of land. What if we think of doing something on that land for the pollinators?" I was at the time reading, completely random in my explorations for writing this book, about some initiatives in Minnesota aimed at creating tax incentives for photovoltaic farms that would plant native flowers between the panels. The idea was to create habitat for pollinators and contribute to their well-being. Florin's answer came swiftly and almost like a surprise: "Tell me more about it. Do some research and let's see what we can do."

This started a journey of discovery, action, and reflection for me that has continued almost 4 years now, with a scope and implications that comprise not only energy transition and pollinators' survival but also transnational policies and international conflicts. At the beginning, I started learning about the pollinator friendly initiatives in the USA. Under the Clean Energy State Alliance, seven states signed legislation to promote pollinator-friendly development in solar farms (Terry, 2020). In my research, I could trace back the initiative to a Minnesota-based collaboration between nongovernmental organizations and research universities at the beginning of this movement. It now comprises a vast alliance of businesses, governmental and nongovernmental agencies, and research linked to the University of Minnesota Extension. In the State of Minnesota, the Board of Water and Soil Resources has successfully developed and deployed a scoring system that outlines detailed standards that a solar farm must meet in order to claim that it provides benefits to gamebirds, songbirds, and pollinators. This assessment is based on a wide range of factors including domination by native species cover, presence of wildflowers, specific diversity in vegetation, available habitat, and other characteristics of the solar sites. Only sites meeting these clearly defined standards are celebrated as being Solar Friendly on the Minnesota Board of Water and Soil Resources' website (Armstrong, 2021).[1]

The benefits for such initiatives are numerous, and research is still in progress (Ghosh, 2023; Marrou et al., 2013; Widmer et al., 2024). I will list a few of them here, as I made the case for the business world.

Benefits for the Environment for Climate Adaptation:

- Land Stewardship
  - Preserves and enhances the land for long-term (50+ years) reuse, stabilizes the soil, and prevents erosion
- Water Management:
  - Offers a better rate of water absorption
- Positive impact on the local ecosystem, biodiversity, and natural capital preservation
- Habitat for Pollinators: Contributes to slowing and reversing the declining of bees and other pollinators populations

Benefits for Financial Institutions:

- Offers a risk mitigation strategy to clients, while protecting the integrity of ecological-based equity incentives
- Offers transparent, measurable, and visible criteria for ESGs to avoid the pitfall of greenwashing
- Directly addresses a new chapter in finance, such as natural capital, biodiversity, and climate adaptation

Benefits for the Solar Farm Owner/Operator:

- Financial Advantages
  - Provides financial facilitation in relationship with financing bodies and medium- and long-term savings

[1] Minnesota Habitat Friendly Solar Program | MN Board of Water, Soil Resources.

  - Increases the eligibility for funding both with financial institutions and at EU fund levels.

- Maintenance

  - Increased energy efficiency, offering natural cooling of the panels
  - Increased lifespan of solar panels due to vegetation keeping a lower temperature at the level of the soil

Benefits for the Community:

- Benefits the agricultural yield from the surrounding crops by increasing presence of pollinators
- Offers a remedy to increasing land competition by deploying multiple land uses
- Provides both a sense of energy security and validation of positive environmental impact
- Creates a "living lab" for scientific communities to study and evaluate climate adaptive solutions
- Offers partnership opportunities with educational institutions by providing field-study opportunities for children and young adults

I packed those ideas with a call for action for one of Florin's clients who aimed to build a photovoltaic farm in a PowerPoint presentation, and in late 2021, I made its case. At the first meeting, I was asked directly: "ok, but what is your business plan?" In other words "how do you make money?" and the interlocutor was thinking about "serious" money, not the modicum that would keep my still nonexistent business afloat. I did not expect this. I thought the arguments of a win-win situation for both pollinators and the business would be enough. I said something I had rehearsed beforehand with Florin that is that we aimed at developing best practice guides that would further help our consulting profile. The question came quickly after: "how much does it cost?" I answered that: for the surface discussed, it would be somewhere around 100,000 euros for the first year, with costs decreasing for the subsequent periods. The answer was trenchant and encouraging: "Oh, that's peanuts! Let's do it." The overall value of their project was 70 million euros. Unfortunately, easier said than done, but let us not anticipate or lose hope.

In February 2022, Russia invaded Ukraine. One of the questions at the forefront of European politics became energy self-reliance and the necessity to decouple Europe from Russia's natural gas exports. The solution envisaged was to accelerate the energy transition, and solar power was one of the strategies. The EU Commission put forth the programme REpowerEU, and the industry started having a lot of wind in its sails. At the same time, the EU Commission put forth a strategy for pollinators. We thought this would be the perfect foundation for our idea. Together with Florin, Valentina Ivan (an energy transition specialist), and Shelby Strand (a leader in education from Canada whose enthusiasm was crucial to start and structure our company), we incorporated a consulting firm under the name Re-Genera Group and contacted the bureau at the EU Commission for pollinators. We were surprised to see that among the initiatives that they encouraged, the pollinator-friendly habitat in photovoltaic farms was not present. We arranged a zoom call and happily presented to them the idea and asked for guidance. The answer was surprisingly cold, and it can be recalled like this: "we only deal here with wild pollinators, not with the bees. It's a good idea, but we don't know what to tell you." We were honestly baffled, particularly at what we perceived to be a very siloed approach on the matter.

Meanwhile in Romania, legislative changes were on the way. Land previously used for agriculture needs to be repurposed in order to develop photovoltaic farms. The implications are complex. The idea itself of taking land from (industrially) producing food created a public dispute that artificially got polarized in terms of food versus agriculture. The dispute is a rather false one, at least in the case of Romania; the surface that would theoretically need to be allocated to photovoltaic farms in order to meet the REpowerEU targets (European Commission, 2022) would represent only 0.5% of Romania's agricultural land. However, rhetorically the polarization sounds good and mobilizes. The unrest of industrial farmers all over Europe is mostly due to the increased prices of diesel and the decrease in subsidies. The solar panels added to the grievances. However, under the urgency of energetic self-reliance, the EU issued a directive to encourage countries to facilitate the repurposing of land from agricultural to photovoltaic farms for surfaces that are less than 50 Ha.

The Minister of Agriculture in Romania at the time interpreted this directive in a particular key: he indeed facilitated the permits for surfaces that were under 50 Ha but blocked any permit issues for surfaces bigger than this. All of a sudden, big projects, such as the one we were managing for our pilot initiative, found themselves at a halt, scrambling for solutions. One of them was to parcel the big surfaces into smaller ones and apply for permits in a fragmented manner. Speculations went on about the inexplicable Minister's move. Was he influenced by the farmers' lobby? In Romania, farmers form a very strong group of influence, obtaining among others yearly derogations from the EU interdiction to use neonicotinoids for increasing production. These substances are very harmful to insects, pollinators included. Was he acting under the influence of the Kremlin, in the context of a war next door? The move seemed to play well in Putin's interest to keep Europe addicted to Russia's gas. No one of course knew, but the effects were visible for everyone. We, for our part, felt that no one had any ears for pollinators and bees. However, we did not lose hope.

In 2022, World Wildlife Fund (WWF) Romania launched "Bee Active!" (WWF Romania, n.d.)—an initiative for pollinators meant to influence policies and practices for a pollinator friendly environment. I contacted them and joined a first meeting in winter 2022, in Brasov, a city 160 km north of Bucharest. I took the train and spent a fruitful day in the company of NGOs and apiculturist organizations, among which was one of the initiators of the Avaaz.org petition mentioned earlier. I was the only representative of a business, everyone else being part of the associative activities. There, I met the director of the program, Monia Martini, with whom I gradually built trust and to whom I explained our project in view of creating alliances across sectors. We decided that it would be great to generate a common initiative that would aim to create a best practice guide and an aspirational standard. In a short while, we created a common logo and started to also look for alternative fundings (see photo). The ambition is to create projects in which the impact of planting local flowers and other small measures would be measured and presented as evidence for an increase in biodiversity.

Meanwhile, in Europe, the banking system entered into a complex reform aiming to take into consideration biodiversity and climate adaptive solutions in their decisions to finance projects. This is done under the pressure of eliminating greenwashing—the practice of claiming sustainability in its absence—and to reform the Environmental Social Governance (ESG) practices. Banks created departments concentrated on biodiversity, and there was a moment in May 2022 in which Deutsche Bank was raided by police and accused of greenwashing (Murray et al., 2022).

Re-Genera's argument was supposed to be simple: The high cost of the initial investment should push investors to seek access to financing at lower costs. This automatically orients them toward meeting Environmental Social Governance (ESG) standards. The intensive use of land and the dilemma it causes (food or energy?) may point to choosing the path of agrivoltaic, that is, using the land under the panels for (intensive) agriculture. While this is an important direction that has the potential to mitigate potential conflicts over land use, it does not necessarily answer ESGs—as intensive agriculture is not necessarily sustainable, and many agricultural lands are in the process of rapid deterioration directly caused by industrial monocultures. ESGs themselves are in a period of crisis and in need of redefinition; too loose criteria had already led to misplaced financing, greenwashing, and lost opportunities (Tricks, 2022).

Land needs restoring and regenerating, and ESGs need to measure less but more efficiently. A way forward to help land regeneration and agriculture all the while contributing to the repair of ESGs is to add substance to marketing, that is, to deploy medium- and long-term strategies for land stewardship that go beyond the simple dual use of land, all the while standardizing them for the industry and finance alike. In conjunction with, or instead of intensive agriculture, photovoltaic projects could devise innovative ways to create pollinator habitat, engaging the scientific and local communities and thus helping the surrounding larger human and natural environment, agriculture included. A standardized set of criteria for building photovoltaic farms, based on scientific approaches on land preservation, accompanied by local and regional policies and incentives to implement them could mitigate the risk of greenwashing (European Commission, n.d.).

Between 2021 and the current moment (early 2024) in Europe, the movement of "greening the green energy" increased. There are a few projects working on a similar logic in Holland and Italy, and the British government developed an indicator of Biodiversity Net Gain (Government UK – Department for Environment, Food & Rural Affairs, 2021), to be used in order to measure the impact of proenvironment measures (Lancaster University, 2021). The builders of photovoltaic farms in Romania, big players on international markets, started being more and more interested; there are a variety of them, from standalone producers of energy to industrial manufacturers that build their own energy supply photovoltaic farm in order to offset carbon on the carbon market. Some of them even became interested in the life of pollinators, but 2023 was the year of securing permits under the new legislation, and (another but) for some, the association with WWF-Romania did not sound. Rumors of corruption and funding received from Russia tainted the image of the international organization. I do not know, and it is not my purpose to find out if this is true. However, these made at least one of the businesses back down. Another one made a calculus: associating with WWF would drive the price of the MW up (by a few cents) and the cost analysis inclined against it. True, this could be also a practical critique of the big overheads that big international NGOs and development industry add to otherwise interesting initiatives. Can they be done for less? Perhaps, and we are trying to find solutions as I write.

However, no one calculated the cost of *not doing* it, at least not from the perspective of the environment and the pollinators. The talks, almost always, returned to quantification: "how much will we save with this?" "Well, studies claim that the natural cooling system of the plants increases productivity by up to 15%". "Can you guarantee this?" Naturally, I could not. Nor could I guarantee the exact quantum of positive impact on the agricultural crops in the vicinity of the photovoltaic panels brought by the proximity of pollinators and their increase in diversity. It seems that the business world has not found a way yet to "value" biodiversity and to guarantee thus its accuracy. In other words, we encounter in this type of interaction the same imperative of bringing a phenomenon (i.e., biodiversity) into the realm of a number-based system of value (finance) and thus to transfer the exactitude of this system into the accuracy of the claim. Sounds familiar? Because it is—this false equivalence born at the end of XVIth Century is the basis of capitalist socio-economic organization.

Entire divisions in big banks are now created to do exactly this: create an algorithm that assigns value to biodiversity. Take, for example, the Biodiversity Fact Sheet published by Commerzbank, one of the most powerful banks in Germany (Commerzbank - Group Sustainability Management, 2023). It is filled with quantification expressed in percentages and contains claims such as "More than 50% of global economic output depends on functioning ecosystem services." One may wonder how the other almost 50% of economic global output functions independent of "ecosystem services." The answer is simple: by conceptually separating humans from "ecosystem," by fragmenting and assigning value, by data-ifying and thus making it easy to digitize the surrounding world, and by modeling within the exactitude of the number based system. "Digitization" of course comes from "digits" meaning "numbers," but perhaps many people do not remember this. The French word for it is perhaps more intuitive: "numerization." With this, one operates and reifies the separation subject/object, keeping agency in the realm of humans and (imaginary) facilitating action through digitizing and measuring. The experts, a socioprofessional category proper to the fragmented world, are its rulers. On top of the hierarchy, perhaps there are the valuation experts. In *The value of a Whale. On the Illusion of Green Capitalism*, Adrienne Buller tackles precisely the matter of assigning value to a disappearing world, in a neocolonial view of nature. She identifies the rise of financialization of biodiversity and the creation of "markets in ecological 'services'" as the new frontier of capital expansion. This very expansion that brings within the logic of the market and risk mitigation categories such as "carbon emissions, biodiversity loss, or the disruption of ecological resilience, from crop failure to disease" (Buller, 2022, p. 264) is based on valuation or numerization of these risks, while hiding behind the numbers the details—most importantly that of context, locations, effects, and interlinks. Thus, for example, the talk of setting aside 30% of Earth for wildlife is not very specific in terms of geographical location and impact on "the other" 70%. Numbers are seductive, but, as in Mueller's story about the salmon industry, they are deceiving particularly because they are abstract and ultimately deny context. The consequences, in the author's view, is the perpetuation of a neo-colonial system

> in its most distilled form: the forcible transfer of sovereignty from the people who occupy the land to those with sufficient monetary and coercive power to insure it is used in their interests. In the process, the inhabitants of these natural capital-rich or service-producing areas -- overwhelmingly Indigenous or subsistence farming communities in 'underdeveloped' regions of the Global South -- are reduced to 'service providers' or custodians for the demands of the global economy (Buller, idem, 265).

This very organization of the narrative of sovereignty is the central piece in valorization, numerization, and/or digitization, because it points to the assignment of agency and subjectivity—thence of sovereignty. Timothy Mitchell, an anthropologist who wrote extensively about colonization, particularly in Egypt, and about the carbon economy, presents explicitly how the political economy of expertise keeps in place the objectified world. In the first chapter from his volume *The Rule of Experts* (Mitchell, 2002) suggestively titled *Can the Mosquito Speak?*, the author brings to our attention the necessity of keeping agency within the human realm for this type of model, such that the management of the objectified world could continue. Mitchell analyzes the events surrounding the building of the Aswan High Dam that included war, a malaria epidemic, and innovations in the chemical industry to move away from the simple explanations based on the centrality of human agency. The combination of factors that form the entirety of the Nile ecosystem, human action included, led to events that caused famine and death at large scale among Egypt's population. The Aswan dam interrupted the Nile's floods that would naturally fertilize the lands and rendered local agriculture dependent on chemical fertilizers. The war caused a shortage in fertilizers because the base substance, the nitrates, were used for explosives. This led to famine. The newly built irrigation channels, constructed to supply water to the crops, and make up for the caging of the river that previously irrigated the lands without "expert" intervention, formed perfect habitat for the *Anopheles* mosquito to move from South to North, where it found a malnourished population, exposed and vulnerable to the disease the mosquito was carrying. The author argues that the heterogeneity of these factors, and the fact that "modern explanations" are wary of linking across fragmented domains and concentrated on human agency and also on a particular type of centrality, do not allow for a more holistic explanation. The problems he identifies are generated by two factors: the human agency centrality and the intention to fit a particular narrative into a more general and universally valid explanation. The universal narrative proposed as generality is in fact very particular and within a very clear Eurocentric perspective:

> Inevitably the generic case in such accounts is the history of Europe or the West, and the particulars of what happened outside Europe are explained as replicas of Europe's history, or variations from that historical pattern, or alternatives to it. In studies of Egypt, for example, events like those I have been describing fit into a variety of larger narratives: the story of the nation and its development, the growth of new social classes and other national actors, and the rise of the modern state, often placed in the context of the development of capitalism, the expansion of Europe, or the global history of modernity. The story takes its shape from the way it fits into a sovereign narrative told about every place, the story of rationalization, technological and social progress, the growth and transformation of production, and the universalization of the culture and power of the West. (Mitchell, 2002, p. 28)

The actors that this story proposes are invariably the humans, with the world as objectified background for their action. The experts are those who drive it. The "quantification of the world "is the technology that sustains this narrative, and AI as it is used today only reinforces it, because it is used within this narrative. The narrative is that of extractive capitalism that perpetuates exploitation of anything deemed "object." It also does something else: its profound analysis would allow us to see the cracks in the system of value and agency assignment. If we are to deconstruct the binaries of subject/object world and accept that agency is hybrid and not necessarily the domain of humans, we would first have to agree that capital itself has agency, a matter that first appeared in Marx's writings. However, while he identified how "the capitalist" ends up embodying "the will of the capital," the subsequent commentators spent too little time in analyzing the consequences of this observation on both the construction of binarisms that separate humans from anything else and humans one from another among themselves. Nor any significant attention was given to the possibility of nonhuman-centered value and agency, not expressed in, but rendered possible by this interpretation:

> Marx, as Derrida says, was 'one of the first thinkers of technics,' the first to grapple with the hybrids of man-machine, capital-consciousness, automatism-will. His writing grasped that human consciousness is an artifactual body, even if in the end he always wanted to ground his critique of consciousness in absolute distinctions between real and abstract, presence and representation, object and value, labor and ideas. These are distinctions whose apparent stability we must explore. How is the ambivalent relation between the nonhuman and the human, or the real and the abstract, constituted? How are the irreducible exchanges or tensions between the two resolved in modern politics into so simple an opposition? (Mitchell, idem)

I attempted up to now in this volume to give partial, fragmented, and hopefully useful answers to these questions. To return to the "cracks," let us think if and how humans—and particular sovereign humans that think themselves as subjects and agents over those who lack agency, be they other humans or nonhumans—may, slowly, move away from the center of the vision and perhaps embody (if there is no other way) the will of something else rather than the capital.

In the presentations I prepared for the potential clients of Re-Genera, I inserted two maps (Fig. 8.1a, b). One is the calendar that explains the alternations of planting, grazing sheep, and transhumance apiculturist that would ideally bring bee families close to the photovoltaic farms, in facilities built for that purpose. The other is a

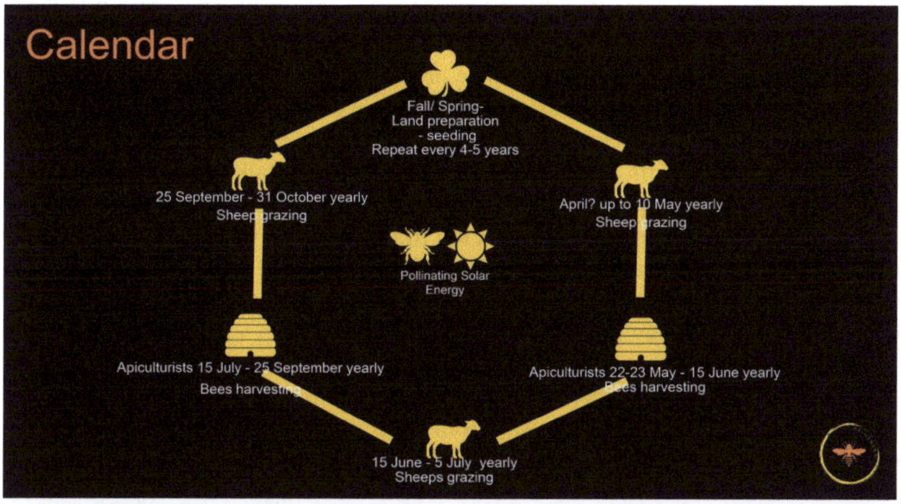

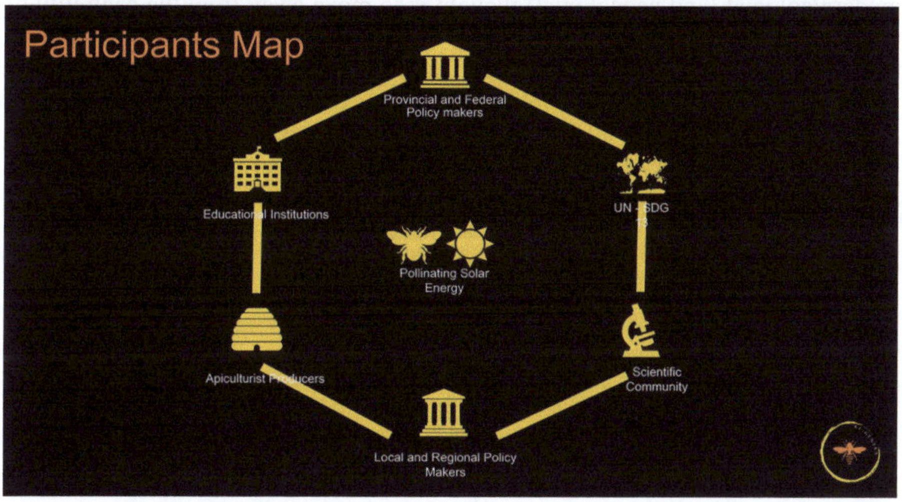

Fig. 8.1 (**a**) Calendar for Re-Genera Group initiatives—by the author. (**b**) Stakeholder map for Re-Genera Group initiatives—by the author

"stakeholder" map—I called this category "Participants." Both have at the center a pollinating insect and the Sun (see photos). My dear colleague from the Royal Roads University, Erin Dixon, who inspired much of the writings here, rightfully pointed out that I should have also put the moon in the center, as the Moon's cycles are crucial for our world. I keep here the first form that was presented to the prospective clients for accuracy, but the moon is there now.

I hoped that this image would suggest moving away from the centrality of human preoccupation to having the pollinators as drivers of the actions. Thinking retrospectively, the discourse, while perhaps visually interesting, was too much for the people who embody capital's will to hear. Hearing "the will of the bees" is not a priority, unless quantifiable. The challenge is the gradual passage to different types of wills. It is a challenge that needs to be brought into all the realms, including the business one, and not kept to be argued about its necessity within the siloed political economy of the expert discourses, because all domains, including that of decapitalization, generate experts that have great difficulties in talking to each other or accepting each other's perspectives. This is what clearly happened when WWF's participation was perceived with suspicion in the business world. The same suspicion, I must add immediately, welcomed me from the part of some WWF employees when I first contacted them. Trust is built in time, and only with patience can we overcome barriers of any kind.

Keeping the narrative of sovereign human in place perpetuates the subject/object separation along with all other binaries and separations that act within and reshape our vision of the world. Moving away from the centrality of the "human subject" in order to understand "human subjectivation" within the complex system would make room for all forms of life to coexist. Indeed, one of the possible ways out of the modern/colonial (and neocolonial) thought is to move away from the centrality of the human subject. This centrality on the one hand places the idea of humanity as the pinnacle subject, and on the other hand it creates racial–/gender–/ethnic–/national–/sexual–/class–/geographical-based categories that hierarchize humans. Bringing in the realm of subjectivity and subjectivation, the "other than human" dissolves these categories and opens up possible futures for the continuum of life, away from compartmentalization. The effects of such a move are far reaching. I would like to end this chapter with an idea developed together with Wanda Krause:

> We cannot afford approaching new problems with old tools, not more than we can afford relying on the centrality of individuality and humanness in leadership theories, if we are to remain in any way human. If we try to continue to minimize the role of collective, of symbiotic assemblages, and of the non-human in the complex systems in which we exist, not only will we not be able to address the challenges that the transforming environment will subject us to, we will also lose our ability to be human. This is simply due to the fact that, while perceiving the necessity of complex and collective leadership, we would try to step into the roles that are not ours to begin with, and we would try to replicate ourselves as (the) non-humans with which we should in fact share the leadership role. Human leadership in this new context may as well mean knowing how to step down and to not attempt to place human agency in all the nodes of the system, allowing the complexity, variance and diversity of agencies to manifest themselves and create the emergent collective and symbiotic leadership. Planetary health requires it. (Krause & Balasescu, 2023: 90)

References

Armstrong, A., Brown, L., Davies, G., Whyatt, J. D., & Potts, S. G. (2021). Honeybee pollination benefits could inform solar park business cases, planning decisions and environmental sustainability targets. *Biological Conservation, 263*, 109332.

Buller, A. (2022). *The value of a Whale: On the illusion of green capitalism* (1st ed.). Manchester University Press.

Ghosh, A. (2023). Nexus between agriculture and photovoltaics (agrivoltaics, agriphotovoltaics) for sustainable development goal: A review. *Solar Energy, 266*, 112146. https://doi.org/10.1016/j.solener.2023.112146

Krause, W., & Balasescu, A. (2023). Towards a truly collective leadership: A case for symbiotic relationships in planetary health and artificial intelligence. In A. Coman & S. Vasilache (Eds.), *Social computing and social media* (HCII 2023. Lecture notes in computer science) (Vol. 14025, pp. 63–74). Springer. https://doi.org/10.1007/978-3-031-35915-6_7

Lancaster University. (2021, October 18). Putting honeybee hives on solar parks could boost the value of UK agriculture [Press release]. *ScienceDaily*. https://www.sciencedaily.com/releases/2021/10/211018172223.htm

Marrou, H., Guilioni, L., Dufour, L., Dupraz, C., & Wery, J. (2013). Microclimate under agrivoltaic systems: Is crop growth rate affected in the partial shade of solar panels? *Agricultural and Forest Meteorology, 177*, 117–132. https://doi.org/10.1016/j.agrformet.2013.04.012

Mitchell, T. (2002). *Rule of experts: Egypt, techno-politics, modernity* (1st ed.). University of California Press.

Murray, M., Kollewe, J., & Foy, L. (2022, May 31). German officials raid Deutsche Bank's DWS over 'greenwashing' claims [News article]. *Reuters*. https://www.reuters.com/business/german-police-raid-deutsche-banks-dws-unit-2022-05-31/

Terry, G. (2020). *State pollinator-friendly solar initiatives* (article). Cleqan Energy States Alliance – CESA. (https://www.cesa.org/resource-library/resource/state-pollinator-friendly-solar-initiatives/)

Trick, H. (2022). *A broken system needs urgent repairs* (special report). in The Economist – ESG Investing - economist.com – https://www.economist.com/special-report/2022/07/21/a-broken-system-needs-urgent-repairs.

Widmer, J., Christ, B., Grenz, J., & Norgrove, L. (2024). Agrivoltaics, a promising new tool for electricity and food production: A systematic review. *Renewable and Sustainable Energy Reviews, 192*, 114277. https://doi.org/10.1016/j.rser.2023.114277

Websites

Commerzbank AG – Group Sustainability Management. (2023, May). Biodiversity factsheet [Fact sheet]. https://www.commerzbank.de/ms/documents/en/factsheet-biodiversity.pdf

European Commission. (2022, May 18). REPowerEU: A plan to rapidly reduce dependence on Russian fossil fuels and fast forward the green transition [Press release]. https://ec.europa.eu/commission/presscorner/detail/en/IP_22_3131

European Commission. (n.d.). Action for pollinators. https://wikis.ec.europa.eu

Government UK – Department for Environment, Food & Rural Affairs. (2021, July 7). Calculate biodiversity value with the statutory biodiversity metric—When and how to measure a habitat or development's impact on biodiversity (guidance) [Updated February 27, 2024]. Retrieved March 1, 2024, from https://www.gov.uk/guidance/biodiversity-metric-calculate-the-biodiversity-net-gain-of-a-project-or-development

World Wild Fund (WWF) Romania. (n.d.). Bee Active—Mobilizarea cetătenilor pentru conservarea albinelor şi a altor polenizatori din Romania [Bee Active! Mobilizing Citizens for the Conservation of Bees and Other Pollinators in Romania]. Retrieved February 2, 2022, from https://wwf.ro/ce-facem/bee-active/

Chapter 9
What Now?

"Each month since June 2023 has seen new temperature highs for the time of year." writes BBC Science (Poynting, 2024), on the very day I am writing these lines, March 7th 2024. On February 14th, 2022, the article *AI-synthesized faces are indistinguishable from real faces and more trustworthy* (Nightingale & Farid, 2022) was published in the PNAS Nexus (Proceedings of the National Academy of Science, USA). The article was submitted in 2021. The title says it all.

In this volume, I attempted to show the intimate and convoluted link between these two types of events. Resumed, this link could be "the alienation of the self from the world." The result is highlighted when reading these news articles together: the neglect of the environment, paralleled with the fixation on the self. Of all things AI could help us with, we fixate on using and refining it to imitate our own images as a conglomerate of pixels on flat screens—literally and metaphorically. I am not saying we do not do all other possible things with it, to a certain degree. I am saying that perhaps, because we are fixating on ourselves, we fail to adopt two major necessary perspectives on AI: first, that this technology only augments ourselves and our place in the world within the way we think of ourselves (and the world), and second, that the dominant worldview or "thinking about ourselves" that forms the playground of this technology is one imbued with values of individualism, competition, and separation from the very environment we pretend we protect. It is worth the question if we are really neglecting the environment, or knowingly and actively destroying it? In other words, are we moving from an egocentric to an ecocentric view or by generalizing an AI-centric view we are about to augment and optimize the egocentric one?

My great-grandmother had another story, one about snakes. She told me that each house has a snake. It is perhaps worth mentioning that the popular name in Romanian for the variety of grass snake *Natrix Natrix* is "house snake." So the house snake resides in the wall or under the basement stones of any house. It is a kind snake, and it is an important snake. She rhythmically hits the wall of the house

A. Balasescu, *Climate Change in the Age of Artificial Intelligence*, Artificial Intelligence, Simulation and Society,
https://doi.org/10.1007/978-3-031-90042-6_9

with her tail, and if you listen carefully, you could hear the beats, like a horologe's tic-tocs or those of a heart. They are the heartbeats of the house, and they mean that the house is alive and the people living in it are okay. If the beat stops, someone will die, and the house will be most probably abandoned, even if it may still continue to stand as an empty shell. This is why we need to take care of the house snake, she told me, and never kill her. In fact, she was saying, do not kill snakes at all. Or any creature, unless needed for food.

I think that all stories have ambiguity and multivalences. These are characteristics "built in" to our ways of presenting and understanding the surrounding world. But we do not always accept them. It is the same with AI, or any technology, as it is with the way we look at snakes or at nature through extension. Can we accept and operate with multivalences and ambiguities? Will we try to use technology in order to emphasize some characteristics to the detriment of others? Command and control over processuality and indeterminacy? Linearity and limitless accumulation vs circularity, redistribution, and perhaps sustainability? For each of these questions, different cultures have different answers, and different fragments within "a culture" may have multiple different approaches.

Within the answers, or based on these, technologies are designed, integrated, and used toward goals that are defined by these answers—and they are always themselves partial, incomplete, and multivalent. Why does Europe have a different approach to AI than the USA and emphasize regulations and ethics over profit and legal normativity? Why is China as an AI Superpower overdiscussed and even more misunderstood, where AI, politics, and society intertwine in ways that Western observers associate with authoritarianism? Why has India recently promulgated a set of policies that requires any AI model to receive state approval before launching into the market and turned into the fiercest state regulator of the AI market and development (Singh, 2024)? Complex answers are provided in many studies and articles that look in detail at the interactions of social and cultural norms, long- and short-term politics, business logic, and historical dynamics. Simply though, the answer is because nothing is deterministic at all levels, and technology is much less than anything. The way we think and use technology is what matters. What does this mean? It means that we humans live in complex systems within which thinking about who and what we are influences the ways we organize ourselves and use our technology which in turn influence the way we think about who and what we are.

Therefore perhaps a new set of questions is necessary in order to create for all of us a place on this planet. When I say all of us, I include every form of life. How do we conceive ourselves, as humans, in relation to the technology we create, and with the environment we populate? When do we perceive them (both technology and nature) as part of us or simply as a complex tool, and when do we endow them, separately or together, with almost mythical character? How do these instances reflect—and get reflected in—the way in which we relate to, use, and abuse both environment and technology? And what does the way technology is conceived of, built, and deployed in relationship with the environment say about us as humans, in our cultural diversity?

I hope that this volume provided some of the answers I propose for these questions. They are not intended to be definitive answers (although some of them may be) but rather propositions and possibilities of thinking that may provide different paths of relating to our own products and tools. They are not intended to reduce ambiguity but on the contrary, and if possible, to embrace it, even with the risk of increasing it. I hope for this book to be read as a plea for releasing definitions and predefining ways of being centered on possessive individualism with identitarian fixations that eliminate the possibility of becoming. I wish that this philosophy, encapsulated in this phrase, would cease to be at the basis of our political economy and at the heart of our technological arms races. Otherwise, we risk using this powerful technology to augment power structures that completely objectify the world, human beings included. What I wish in fact is to join my voice to the chorus of those who, in many ways and under many names, advocate for a dynamic and relational understanding, away from a mechanistic view, in which the individual has its place within the web of relations and not at the center of it. It is also a view in which the individual is not bound to transversal, temporary identification projected as longitudinal and deterministic identity, that is, a view of beings in constant becoming. So the question is, how can we integrate AI into this view? Certainly not by wanting or trying to make the world fit the AI models. Here probably it is time to revisit the examples in the opening chapters. The workarounds healthcare providers in Thailand and in Canada devised in order to deal with concrete situations outside the lab should be in fact models of adaptation of technology to the field conditions and not considered failures to "correctly" use it as suggested by the very word "workaround."

There is perhaps a story of a mirrored image showing the extremes of the polarity and ambiguity in approaching technology in Japanese restaurant practices. On the one hand, we have a very technofriendly type of restaurant with robot waiters. Everything goes perfectly and smoothly. The robots take the orders and bring the food, never misunderstand the orders, and serve the clients efficiently and promptly. At least this is the story. On the other hand, there is for the moment one restaurant in Japan that hires elderly people as waiters, who suffer from dementia. The idea is to keep these persons integrated into the social system, which in itself is a manner of creating well-being for them and the community. The clients come and they are aware of this. More often than not orders are mixed, wrong food and beverages are brought to the tables, but no one is getting upset. On the contrary, whatever the waiters bring is accepted, and the ambiance is one of general amusement and laughter.

Hearing these two restaurant stories, one would be tempted to say that they each have their meaning and that they each serve different expectations. That is true. However, what the first story omits to take into consideration is the following detail: in robot-waited restaurants, there is one person whose job is to take the food from the robot tray and put it on the tables. The robots cannot do it by themselves; it is too complex an operation. This image alone, of a human serving the robot waiter's needs, should make one revise and reconsider the idea of hypertechnologization and to ask again: *what are we trying to achieve?*

Personally, I consider that food eaten in a relaxed ambiance with laughter and accepted mistakes contributes much more to the well-being of both persons and communities than that "perfectly" served by a robot assisted by a dehumanized pair of hands. This image is very much reminiscent of Henry Ford's infamous exclamation: "Why is it that every time I ask for a pair of hands, they come with a brain attached?" The dream of detached arms is perhaps the dream of technology as integrated into this particular philosophy of measurement, efficiency, and endless optimization. This is very close to the image of eternal peace in the technological "Kingdom come"... The technooptimist discourses on artificial intelligence contain, if pushed to their ultimate logic, the "dark side" of Fordism, already at work in the roborestaurants: replace the human "imperfect, subjective, emotional, and always already decaying" brain with the "perfect, objective, rational, and immortal" AI and use only the perfect dexterity of human hands, yet irreplaceable—but who knows in the future? Perhaps it is time to be reminded, as Warnier puts it, that humans think with their fingers. And so do octopuses. And also that thinking is done collectively, in greater systems, as an emergent quality and not as an individualistic quantity of output.

The challenge today is complex and ambiguous: to understand how a certain system of beliefs influences our creation and views on AI and the environment; how this system is perpetuated, augmented, reinforced, and reorganized by the very use we give to AI; where does it take us; how does it changes us; how can we change it by changing the angle of the story we tell ourselves about AI and the environment; and to what extent?

By looking again, at the end of this path, at the triad of nature, culture, and technology, we can understand that in fact it is not necessarily a triad, although we may think of it this way. The level of separation or integration of these elements depends on the level of the analysis under which one places oneself. In an article published 7 years ago, together with my friend and colleague Thomas Seguin, we proposed a scheme of understanding this tripartite construction and its role in the processes of subjectivation as a holistic, relationally structured system. We also attempted to explain change and how change can emerge from an accumulation of surplus in any of the system's relations, not as a causal linear phenomenon but as a circular one. We called this a model of energetic transfer and subjectivation (Fig. 9.1). Like any model, it is wrong. But it may be useful. I reproduce the scheme here, with a brief explanation:

> Ours is not an argument of causality, but one based on circularity and apparently spontaneous occurrences. This kind of circularity between entities and disruption within a whole makes us think of a structure. In that frame, *we can understand structure as an anthroposocial configuration which consists of stabilization of energetical potentials (within matter, mind/body, groups) through routinization of usages and practices, via the technical universe and symbolic mediation.* Change can be generated at any of the levels of interactions discussed above, through different means: an excessive accumulation, a reconfiguration of relationships and interactions, a translation of energetical abstraction, or a combination of those; in other words, a generation of surplus that would in fact generate the change. (Bălăşescu & Seguin, 2017, p. 261)

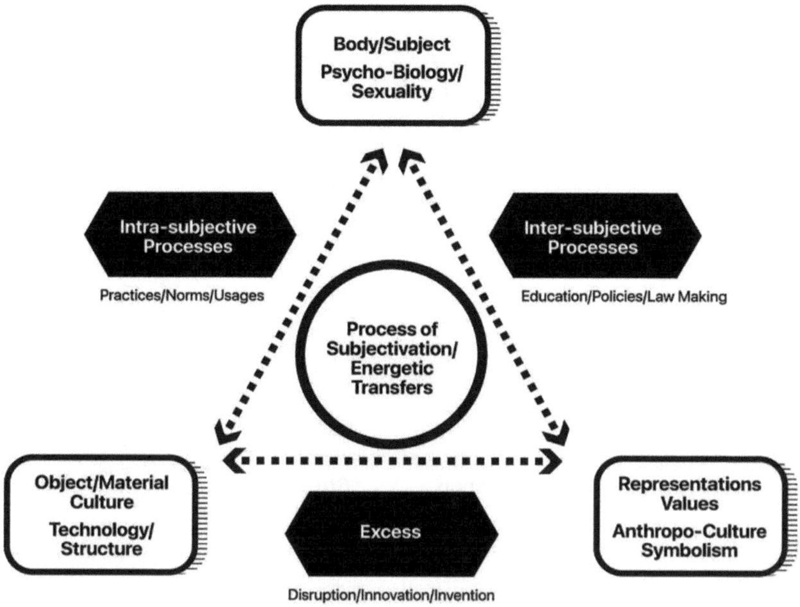

Fig. 9.1 Model of energetic transfer and subjectivation. Graphics: Bojan Spasic

At the center of this representation is not the individual, the self, but the process, the transfers, and the relations that continuously contribute to forming subjectivities and many other things. But the emphasis is on the dynamic interrelationships. A broken down representation of this graphic shows each point of the triangle as a triangular mechanism itself, mechanisms within mechanisms, and relational and potentially infinite (Fig. 9.2). The center, the subject itself, is ontologically empty; it is being perpetually constructed and reimagined through relationships at multiple levels.

The exact opposite emphasis seems to dominate our approach to AI. One of the central issues with the current beliefs surrounding AI is that of the centrality of the individual as a bounded entity with limitless capacity of improvement, including the potentiality to become immortal. There are many things that derive from this vision, from AI design and application to aspirations and expectations of what technology can do, and for whom. Under this condition, AI will only perpetuate, augment, and reinforce this vision of self and individuality. However, there are all other competing views, competing stories that are told with different voices, on different levels. The relational personhood (that cannot be properly called "individual") is central to these stories. We can find declinations of this story in different sciences, from health to physics, in sociopolitical movements that emphasize the necessity of moving away from extractive capitalism, in feminism, and in the reinvigoration of Indigenous and Local Knowledges. For the skeptics, in the past 40 years, both the medical establishment view and the popular representations of the immune system changed

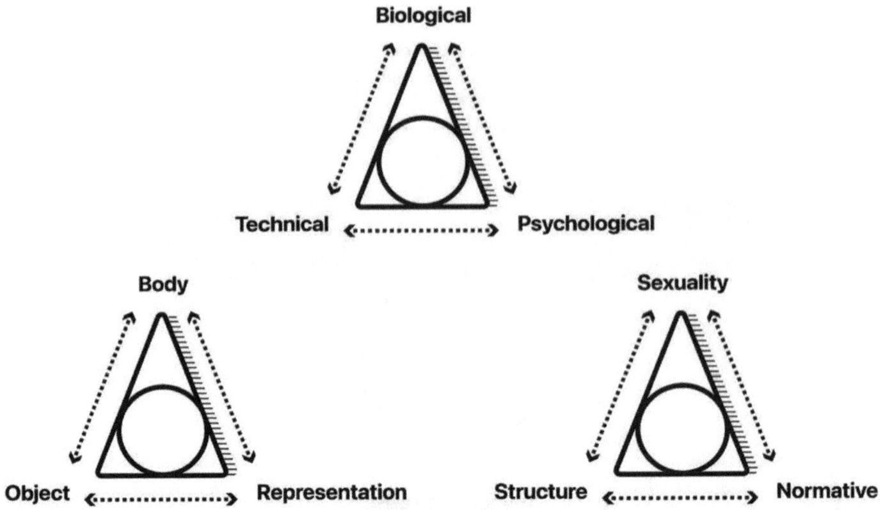

Fig. 9.2 The deconstruction of the model of energetic transfer and subjectivation, Graphics: Bojan Spasic

from one that conceives it as the guardian of a bounded castle (the body/the individual) to a flexible ever adapting and porous continuum of multiple organisms interacting. At first, this happened within the spaces of metaphorical affordance created by the rhetoric of flexibility and economy that the neoliberal doctrine generated in the 1980s and 1990s, as Emily Martin (1995) shows in her historical exploration of this theme. Since the publication of her work on "flexible bodies," the metaphor of flexibility proliferated and created an entire new economy. At the same time, and perhaps paradoxically, it made it possible to think in "other-than individual" terms, contrary to those favored by the very neoliberal rhetoric that emerged from it.

The same type of paradox and ambiguity is at the heart of thinking around AI. While it is indeed created and used mainly within the limited framework of rationality, individual personhood, and capitalist optimization, it also created the possibilities of rethinking relationality and of rejoining other-than-modern ways of being, doing, becoming, and decentering of the human while at the same time replacing it into the web of interconnections at both micro- and macrolevels.

The argument here is that, if we are to survive, we need to embrace this ambiguity and to accept a various degree of relevance of any category we operate with, based on the multilayered aspects of whatever reality we place ourselves in. That is, rationality works at some level, but it is not universally transferable and it has a variety of declinations. Individual personhood is relevant, and it is not always so. Objectivity may work with objects, but not everything in our universe is an object. In fact, very few things are mostly, and perhaps only, what we materially created, to some level of analysis, of course. Mathematics is an exact language, but it is not *the language* that could accurately describe the universe. In fact, we probably lack that language, since we are prisoners of the ones we have. Or that language is always spoken to us, in us, only if we start listening. We certainly should not think that there is only one way of being and doing that is mathematically codifiable and universally transferable. The danger with AI is not that it will "take over" humanity but that it will augment a reduced vision of humanity that takes over everything else while, and by, transforming that everything into objects of exploitation.

When I was in the third grade, my cousin handed me a book, simply called *Frogs, Snakes, Lizards* (Fuhn, 1968), the animals that a lot of people frown upon. The same book was regifted to me after four decades by someone I hold dear. It is a book written by a Romanian herpetologist in the 1960s, and it provides a poetic and sympathetic view on the fascinating world of reptiles and amphibians. It is also a manifest for an integrative view on nature, while being a plea for the marginalized universe of slithers. It definitely contributed to my path in life in ways I am not entirely capable of expressing or understanding for that matter. Here is a passage with a strong metaphorical value:

> [...] I came to love the world of slithers, a world full of surprises and rich in most successful adaptations, created over the millenia by the active forces of evolution. This world can hardly be guessed, if at all, by the man with his raised bat, disfigured by hatred, who shatters it in a rain of senseless strikes. (I.E. Fuhn, 1968, p. 16, my translation)

More often than not, at least in the "modernized" world or in those worlds infused with a negative image of this animal, people kill snakes on sight. The problem with this, besides killing innocent beings, is that it is a general action applied without context. It is a model of algorithmic behavior: if a snake, then kill. Yes, *some* snakes behave dangerously in *some* contexts; and the multitude of possible interactions is possible to be discerned and codified to *some* extent, while context is harder, if not impossible, to quantify. Killing a snake on the spot in order to avoid the *potential* harm that it *might* do sometime in a more or less *probable* future to a person is a decision that needs to be weighed carefully. At some level, the reflex killing of this kind expresses the difficulty of coming to terms with one's own mortality. And it sounds to me like *Minority Report*, the science fiction movie depicting a society that arrests and incarcerates people based on predictions of possible future crimes. Can this be the mental model that informs some beliefs we hold about AI and that the perfect AI-driven societies may drive us to? And this brings me back to my great-grandmother story of the house snake. As her story goes, we should refrain from killing. If we kill the house snake that would stop the heartbeat of our home, and it would leave it empty: the house will stand as a structure, but the home will be deserted by its inhabitants. What if the snake in her story is nature herself with her rhythmic heartbeat that brings life to our house? Are we about to empty it and then perhaps desert it to migrate to Mars enacting the technological fantasies of the Silicon Valley's billionaires? If our current stories about technology as objective tools for command and control, and about nature/environment as a resource that we can technologically dominate lead us to killing, perhaps it is high time to change them. Everything else will naturally help us, including AI.

Remember the seemingly separated realms of nature, culture, and technology? There is a message of hope in the harmonization of those elements that can be a solid tripod for the continuation of life on Earth, as much as they can be the recipe for catastrophe. The choice is ours to make. We better keep it ours.

References

Bălăşescu, A., & Seguin, T. (2017). Another economy: Towards a cultural dialectics between energy and society. *Innovation: The European Journal of Social Science Research, 31*(3), 251–277.

Fuhn, I. E. (1968). *Broaşte, şerpi, şopârle (Frogs, Snakes, Lizards)*. Editura Ştiinţifică.

Martin, E. (1995). *Flexible bodies: The role of immunity in American culture from the days of polio to the age of AIDS* (Revised ed.).

Nightingale, S. J., & Farid, H. (2022, February 14). AI-synthesized faces are indistinguishable from real faces and more trustworthy. *PNAS Nexus*. https://www.pnas.org/doi/full/10.1073/pnas.2120481119

Poynting, M. (2024, March 7). More climate records fall in world's warmest February. *BBC Science*. https://www.bbc.com/news/science-environment-68428348

Singh, M. (2024, March 4). India reverses AI stance, requires government approval for model launches. *Tech Crunch*. https://techcrunch.com/2024/03/03/india-reverses-ai-stance-requires-government-approval-for-model-launches/?guccounter=2

MIX
Papier aus verantwortungsvollen Quellen
Paper from responsible sources
FSC® C105338

If you have any concerns about our products,
you can contact us on
ProductSafety@springernature.com

In case Publisher is established outside the EU,
the EU authorized representative is:
Springer Nature Customer Service Center GmbH
Europaplatz 3, 69115 Heidelberg, Germany

Printed by Libri Plureos GmbH
in Hamburg, Germany